Ant Architecture

Ant Architecture

THE WONDER, BEAUTY, AND SCIENCE
OF UNDERGROUND NESTS

Walter R. Tschinkel

PRINCETON UNIVERSITY PRESS

PRINCETON AND OXFORD

Published by Princeton University Press
41 William Street, Princeton, New Jersey 08540
6 Oxford Street, Woodstock, Oxfordshire OX20 1TR

press.princeton.edu

Library of Congress Cataloging-in-Publication Data
Names: Tschinkel, Walter R. (Walter Reinhart), 1940– author.
Title: Ant architecture : the wonder, beauty, and science of underground nests /
Walter R. Tschinkel.
Description: Princeton, New Jersey : Princeton University Press, 2021. |
Includes bibliographical references and index.
Identifiers: LCCN 2020040060 (print) | LCCN 2020040061 (ebook) |
ISBN 9780691179315 (hardback) | ISBN 9780691218496 (ebook)
Subjects: LCSH: Ants—Nests. | Underground architecture.
Classification: LCC QL568.F7 T578 2021 (print) | LCC QL568.F7 (ebook) |
DDC 595.79/61564—dc23
LC record available at https://lccn.loc.gov/2020040060
LC ebook record available at https://lccn.loc.gov/2020040061

British Library Cataloging-in-Publication Data is available

Editorial: Alison Kalett and Whitney Rauenhorst
Production Editorial: Mark Bellis
Text Design: C. Alvarez-Gaffin
Jacket Design: Sukutangan
Production: Jacqueline Poirier
Publicity: Sara Henning-Stout and Kate Farquhar-Thomson
Copyeditor: Laurel Anderton

Jacket art by the author

This book has been composed in Perpetua Std

Printed on acid-free paper. ∞

Printed in China

1   3   5   7   9   10   8   6   4   2

For Vicki
my nestmate for life

CONTENTS

Preface   ix

CHAPTER 1   Soil, Ants, and Life Underground   1

CHAPTER 2   Revealing the Shape of Empty Spaces   11

CHAPTER 3   Meet the Architects   28

CHAPTER 4   An Inventory of Ant Nests   48

CHAPTER 5   Moving House   69

CHAPTER 6   A Diversity of Architectural Plans   96

CHAPTER 7   Ants and Soils   115

CHAPTER 8   Division of Labor and the Superorganism   135

CHAPTER 9   The Evolution of Nest Building by Ants   175

CHAPTER 10   Afterword: The Future   206

Selected References   215

Index   221

PREFACE

If you wanted to be an explorer ferreting out the undiscovered wonders of North America, you would be way too late. Sure, there might be small patches left to discover, but even these are laid out for all to see on Google Earth, identifiable by their precise latitude and longitude so that your GPS can lead you directly to them. But there is a way in which even today it is possible to be an explorer. It is not geographic exploration, but its intellectual analogue—scientific exploration of the natural world, and it is this exploration that is the story of this book. It has been my fortune to have lived a scientific career of rambling exploration, a career I seem to have been destined for. My attraction to the natural world (and food) was already obvious by the time I was two years old—a photograph shows me with a bunch of flowers in one hand and a piece of cake in the other. By the time I was six, I was telling people who asked that I was going to be a biologist (whatever I thought that meant). Still, the road to my eventual career was not entirely linear, nor was it a carpeted path. I endured an abysmal biology course in high school, most of which consisted of copying definitions out of the textbook and listening to the teacher denounce evolution as devil talk that contradicted the Bible. My chemistry course was even worse, taught by a part-time cotton farmer and part-time holy roller preacher who practiced his sermons on the class. He knew precious little about chemistry. Somehow, my friend BB and I wheedled permission to work in the stockroom while the sermons were going on. We were interested mostly in making dangerous and explosive stuff, and once we almost succeeded in blowing up the stockroom. Another high point was when we ignited a batch of thermite, creating a cascade of white-hot molten iron that burned a permanent star shape on the wooden floor.

After high school, I attended a small men's college ("for small men," we joked), where I was finally able to give free rein to a wide range of biological pursuits and to find like-minded students who shared my interests. At that time, the great and exciting strides in biology were being made in cell physiology and biochemistry. Thus I applied to the graduate program in the Biochemistry Department at the University of California, Berkeley, only to be turned down. When, a week later, I was awarded a National Science Foundation Graduate Fellowship that paid full tuition, I called up the Biochemistry Department and asked whether they would admit me now that I had money. Sure, they said, we would love to have you. So I said forget it and applied to the Bacteriology Department, which admitted me right away.

Biochemistry and bacteriology are pretty similar, and once I got into the practical, laboratory part of the discipline, I gradually realized that, for me, something was missing. Where were the critters that had seduced me into biology in the first place? Where were the beautiful living things with wings, legs, leaves, colors, and forms? Could I really spend the rest of my life looking at flocculent white precipitates in a test tube? There would be a huge piece missing from my life if I continued in this field. I needed to find my way back to what had charmed me about the natural world—the creatures I could see, touch, feel, and handle, whose construction and function were beautiful.

Looking back, I can see that I was trying to decide at what scale biology held the most charm for me. At the time, I saw it only as what kind of objects I wanted to deal with, and where I wanted to do it. Flocculent precipitates had convinced me that my preferred scale wasn't the molecular or biochemical. This was a world of white debris, gelatinous slime, blurred bands on gels, or clear solutions that absorbed ultraviolet light. There was nothing to grasp, no form to appreciate, no movement, no behavior, not even color unless you worked on hemoglobin or vitamin B_{12}. No one could actually see an enzyme violently clutch a substrate and wrench it and twist it until it changed chemically. The exciting things that went on in this world were constructs of the mind, built from elaborate physical and chemical measurements.

My friendship with John Doyen, a graduate student in entomology, showed me a way to return to working with living creatures, not just cell-free extracts of them. John's specialty was tenebrionid beetles, a family of mostly black beetles (from "tenebrous," meaning dark) with many species in the western United States. Particularly appealing was that most species produced a noxious defensive secretion that stained your fingers brown, and some of them could even spray it for distances of up to a meter, and one of them (the mealworm beetle) had even been shown to secrete a sex pheromone. Here, then, was a way to roam the western countryside to collect black beetles from under stones, collect their stinky secretion, and analyze it with the latest fancy, expensive techniques—gas chromatography, nuclear magnetic resonance, infrared spectrometry, and ultraviolet absorption. I had found a way to apply my background in organic chemistry and biochemistry, and in addition, I got to watch the defensive behavior of the beetles and dissect hundreds of them to study and draw the structure of their beautiful glands. With respect to the sex pheromone of the mealworm beetle, I figured out how to make males copulate with glass rods treated with female pheromone. What could be more fun?

At the same time, I discovered the pleasures of building contraptions that aid research. I had imported a large tenebrionid beetle from Costa Rica, but the larvae in the culture refused to pupate. Seeking a cause, I eventually showed that it was the crowding in my culture that inhibited pupation, but was the inhibition chemical, mechanical, or the result of some other process? In order to test the role of larva-upon-larva touch, my major professor and I built a gimmick in which slowly rotating petri dish lids dragged bathroom stopper chain over a single larva in each dish. In the controls, the stopper chain was too short to contact the larvae. The tickled larvae failed to pupate, while the untickled ones pupated quickly. This contraption, nicknamed "The Stimulatorium," was the first of many contraptions that gave me a lot of pleasure. Solving problems creatively and simply had revealed itself to be one of the pleasures of research (and of life beyond research).

Building contraptions did not arise from my research needs but was already part of my life. I have always liked doing things with my hands, whether building shelters or carving wood in the Boy Scouts, repairing my 1946 Ford V8 convertible, building fine furniture, or constructing palm-thatched huts. Contraptions and gizmos were really an application of this tendency, but applied to my research, they served it well. Each contraption was designed to solve a problem or answer a question, thus aiding my research. During the early years of my research, my equipment could be purchased for thousands of dollars, but I made later contraptions in my garage from wood, scraps, plastic, and salvage.

These were the traits, preferences, and abilities that set me on my particular path of biological exploration, beginning with tenebrionid beetles and leading eventually to a wide interest in the natural history of ants. Almost by accident, I began my study of the subterranean architecture of ant nests, eventually broadening my quest from simply making casts of the nests of multiple species to wondering how superorganisms construct nests, and why these have such a large range of forms. As I later learned, the piles of dirt on the surface did not even hint at what lay below.

This book is about my quest to discover the nests that lay below these piles of dirt, how I studied these nests, how the ants constructed them, how they served the needs of the colonies, and how they differed among ant species. Although many ants also construct aboveground nests from soil, litter, or plant material, the focus of this book is only on the nests ants excavate underground.

ACKNOWLEDGMENTS

Many people have been involved in the two decades of my nest architecture studies. Their roles ranged from those who were simply curious to see how I made nest casts to those who helped in important ways. I enjoyed showing off my metal-pouring skills to the former and am grateful to the latter for their help in pouring and digging. Special thanks are due to several former students and assistants: Kevin Haight, Christina Kwapich, Joshua King, Tyler Murdock, Kristina Laskis, Elliott Royce, Henry Tschinkel, Daniel "Julio" Dominguez, Nicholas Hanley, and Dennis Howard. In addition, Kevin Haight, Nicholas Hanley, Daniel "Julio" Dominguez, Tyler Murdock, and Neal George provided competent assistance with several lengthy associated projects. Sandy Heath and Ralph Anderson in the biology machine shop cheerfully fashioned multiple crucibles from steel scuba tanks and cut many aluminum scuba tanks into pieces that would fit into my crucibles. Henry Tschinkel and Dennis Howard acted as videographers and photographers. I am grateful to Jack Rink and Jim Dunlap for suggesting our collaborative project on harvester ant bioturbation. I am especially grateful to Christina Kwapich, Joshua King, Dennis Howard, and my wife, Victoria Tschinkel, for reading versions of the manuscript and providing helpful suggestions and comments. My nest architecture work was supported for seven years by the National Science Foundation.

Ant Architecture

CHAPTER 1

Soil, Ants, and Life Underground

Under our feet lies a mysterious invisible realm. Heaps of soil in the shape of craters, mounds, or strewn pellets (fig. 1.1) hint at its existence. Although many creatures burrow in soil, most of this soil is brought up from below by ants during the excavation of their nests. Ant-made soil piles occur in a wide range of habitats and locations, from the rain forests of Uganda to the sidewalks of Los Angeles (to the degree that these sidewalks exist). Because ants vary enormously in body and colony size as well as in nesting habits, these deposits range from almost invisible to the obvious mounds of fire ants or Allegheny mound-building ants, or the colossal excavations of the leafcutter ants of tropical America, which can occupy as much belowground volume as a modest-sized house.

Fig. 1.1. A soil dump resulting from the excavation of a nest below. The generally crater-like form is typical of many ants, but far from all. Note the US dime for scale. This crater was formed by *Dorymyrmex bureni*. Author's photo.

The excavated soil tells little about the nest below. Conceivably, it suggests whether the nest is large or small, but rain, wind, and animals scatter soil piles, so even this deduction is unreliable. Nothing about the shape of the cavities, their arrangement in space, their depth, or their size is revealed by the excavated soil. Do the nests have a consistent architecture? Is there variation among ant species? How quickly do the ants create these nests? How do they use the space they create? These mysteries may not motivate many people into action, but to me they sound a strong call. What are the ants creating underground, and how does it serve them in their lives?

I am not the first biologist to ask such questions. Most of my predecessors have approached the challenge of revealing ant nest structure by first excavating nests in a range of soils and then publishing their findings as sketches or drawings of longitudinal or cross sections, or serial vignettes of nests (fig. 1.2). Some of these are crude sketches, some are more informative, and a few are excellent scale drawings (for example, fig. 1.2). Most of these were incidental to other studies—as far as I know, few were motivated primarily by a desire to describe the subterranean nest architecture. But all together, these studies give us a sort of "preview of coming attractions" that suggests that the study of ant nests and their role in ant biology might be very rewarding.

I began studying the mysteries of ant nest architecture almost unintentionally a couple of decades ago as a side project of my "regular" research. As I dabbled in this subject, I was increasingly drawn into revealing these mysteries as the main focus of my research. This book is mostly about my own exploration of the underground world of ants, based on the successes and failures of my ant research in the coastal plain forests of northern Florida over the last 25 years. Far beyond merely describing ant nests, I have approached the subject broadly, integrating nest architecture with relevant bits of physics, a touch of chemistry, some soil science, ant behavior, colony biology, ant ecology, ant natural history, some experimentation, and occasional personal adventures and ruminations. I hope to show the reader the attractions, problems, and rewards of pursuing a research subject with a passionate curiosity and a love of solving problems. Indeed, I have always found an aesthetic pleasure in working with

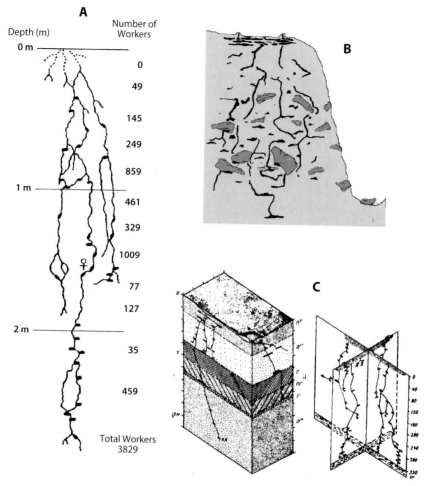

Depth (m)

A

Number of Workers

0 m

0

49

145

249

859

1 m

461

329

1009

♀

77

127

2 m

35

459

Total Workers
3829

B

C

FIG. 1.2. Examples of published drawings of subterranean ant nests, two with scales. Rendering the three-dimensional nature of such nests with drawings is difficult. *A*, from Kondoh (1968); *B*, from Talbot (1964); *C*, from Dlussky (1981).

the "objects of nature" rather than the abstract concepts that are so fashionable (and admittedly important) in modern biology. I believe the reader will find aesthetic pleasure in these objects, too, and will be charmed by the lives of the ants that create them.

I think of myself as a sort of pioneer, mapping and describing an unknown land. This is because biology always begins with a description, and it is no surprise that the infant field of ant nest architecture should begin with a description before taking up a range of brainy hypotheses to explain the observations. It is also probably no surprise that progress in research depends on having available or inventing the methods needed for answering the relevant questions. Throughout the history of science, various fields have blossomed as a result of the invention of a new instrument or process, be it the microscope, the microtome, or any number of other inventions. The field of ant nest architecture is no exception, even though the methods are simpler and more mundane than a synchrotron, a nuclear magnetic resonance instrument, or a confocal microscope. A remarkable amount of interesting stuff can be learned with shovels, plastic bags, a modest ability to count, and a homemade kiln. In an era of high-tech science, I offer a story about the pleasures of low-tech shoestring science.

THE ANTS

The creators of this mysterious underground realm are the ants. In my experience, most people are aware of ants, those pesky creatures that mob the spilled sweet drink on their kitchen counter or make dirt piles on their pristine lawns, but few are aware that the world of ants is like another universe, an alien world. It thus seems prudent to start with a brief sketch of ant biology and diversity.

Ants are social insects whose ancestors diverged from the ancestors of wasps some 100 to 140 million years ago. Their societies (usually colonies) are distinguished by a strong division of function among individuals, such that only one or a few individuals are capable of laying fertilized eggs (the queen or queens), while most of the others are more or less sterile and carry out most of the work (the workers). All of the individuals with a social function are females. Males are produced only for mating with queens and are usually present for only weeks out of the year. Typically, a colony is a family whose mother is the queen and whose daughters are the workers. Daughters are full sisters if their mother mated with a single male, and half sisters if she mated with multiple males. At the individual level, ants are typical of insects with a complete metamorphosis, developing from egg to larva to pupa to adult. Sociality has built on this basic insect plan by affecting how ants develop into adulthood, producing either sterile workers or adults with fully developed sexual organs that are capable of mating and reproducing.

Sociality has made the ants an enormously successful group of animals, dominating many ecosystems in most of the warmer parts of the world. Their biomass—that is, their total weight—often exceeds that of any other animal group in their habitat. About 14,000 species have been described, but at the rate of discovery of new species, it is likely that the final count will be 20,000 to 40,000. For example, in his exhaustive sampling of the ecosystems of Madagascar, my colleague Brian Fisher has personally discovered and named over 1,000 new ant species. When queried about the number of ant species, ant experts usually estimate between 20,000 and 30,000, reasoning that much of the world remains poorly explored for ants and other insects. Many of these new species probably already reside in museums, waiting to be described by ant taxonomists, who, unlike the ants, are in short supply.

With their diversity and abundance, it is not surprising that ants occupy a wide range of habitats. Many species are scavengers and predators; some are herders of livestock such as aphids, mealybugs, and scales; still others are specialized predators of such tidbits as spider eggs, or of difficult prey ranging from hairy millipedes to springtails; some are communal nomadic hunters settling temporarily in camps; some gather wild seed crops; and some farm fungus on beds of caterpillar droppings or leaf fragments. Here in the coastal plain pine forest of Florida (where I do much of my research), it is common for all of these lifestyles to be represented in a plot as small as a medium-sized suburban lot.

This wealth of ant species is not evenly distributed on the earth. Rather, the number of ant species by region declines with increasing distance from the equator. Tropical regions, especially in the humid tropics, host between 4,000 and 6,000 ant species, but away from the equator this drops rapidly until at latitudes greater than 50° north or south, there are fewer than 50 species. I once collected a sample of *Leptothorax muscorum* at almost 70° N latitude on the Arctic slope of Alaska north of the Brooks Range. Only two ant species in the Arctic region extend from North America across Siberia, and the colony I found was nesting in a rare sandy bank facing south, soaking up every calorie of sunshine it could get, perched as it was only a few centimeters above the permafrost. In winter, the nest and everything in it froze solid—a life on hold, not to be resumed until the spring thaw. Its life could define the word "tenuous."

Of course, larger areas have more ant species, so equivalent-sized countries must be compared. Ecuador and Finland are not very different in size, but Ecuador has over 700 ant species, while Finland is home to only 64 species. An area of 16 hectares in the Peruvian Amazon yielded almost 500 species of ants. Brazil and the United States are pretty similar in size, yet the United States has only about 800 species while Brazil has over 1,400, and once Brazil is fully explored, it will probably yield many more.

Myrmecologists have speculated and argued for decades about which group of insects gave rise to the ants. For a long time, the consensus was that the closest relative of the ants was a wasplike creature similar to modern tiphiid wasps. These wasps seek out the larvae of beetles, paralyze them with a sting, and then lay an egg on them. The larva hatching from this egg then grows and develops by consuming the beetle larva. More recently, molecular methods have been used to determine the degree of relatedness of various insect groups and to arrange them into family trees (phylogenetic trees). Basically, the sequence of base pairs in the DNA of groups of organisms changes with time, so the number of pairs in a long DNA sequence that differ is a measure of both the time (more or less) the two lines have evolved separately, and the degree to which they are (or are not) related. Recent studies of many families of ants, bees, and wasps have shown that ants are most closely related to bees and stinging wasps. In the language of taxonomists, they are "sister groups." Bees, of course, collect and feed on pollen, whereas ants (at least the more primitive ants) and most wasps are carnivorous or parasitic. However, what ants, most bees, and most stinging wasps have in common is that they build or find nests and bring "stuff" (pollen, prey, nest material) back to the nest. Wasps outside this group tend to find their prey and parasitize it in place rather than haul it back to a central place. The collection of "stuff" is probably the part of the life history and behavior that predisposes these sister groups to evolve sociality, because it makes parents and offspring more likely to associate in the nest. Given other favorable factors, this group can evolve sociality. It is notable that sociality has evolved six to eight separate times in the bees and once in the ancestral ant, all of whose descendants are social.

Nest construction thus facilitated the evolution of sociality, because living in a nest and returning "stuff" to it predisposes insects to form related groups and thus evolve cooperative behavior. In other words, nesting behavior predates the evolution of both ants and sociality. Modern social insects share a particular set of life history features—that is, characteristics of the stages and phases of their life cycle—and ways in which these meet the life challenges of the species: (1) daughters remain in the nest with their mother and sisters (generations overlap); (2) sisters care for younger sisters rather than their own offspring (communal brood care) and coincidentally care for their mother, too; (3) some individuals specialize in laying eggs, and others in rearing them and doing other necessary work (division of labor or function). It is easy to imagine that a mother ant, bee, or wasp that repeatedly hauls "stuff" back to the nest to feed to her developing offspring would eventually share that nest with her adult daughters, and they with their younger and contemporary sisters. The presence of a brood would trigger the brood-tending behavior of the sisters, and various nutritional and perhaps hormonal conditions would suppress the reproductive capability of the daughters. Should such a combination of a mother (queen) and her suppressed daughters (workers) be more successful in producing the next generation than the daughters trying it on their own, then this nascent sociality would be favored by natural selection, and social evolution would be on its way. Such social evolution eventually passes a point of no return when the workers are no longer capable of being queens, as happened in the ants. After that point, ants were no longer able to ask, "Should I be social or should I go it alone?" They were irreversibly committed.

In addition to nest construction and stuff collection, a particular mode of sex determination in these hymenopterans predisposes them to become social. Whereas females develop from fertilized eggs like most animals, males develop from unfertilized eggs. This means that females are diploid (have two sets of chromosomes), and males are haploid (have one set). All of the male's sperm carries the same set of genes. If the queen mates with a single male, math tells us that sisters are 75% related to one another (i.e., 75% of their genes are the same), whereas they are only 25% related to their brothers. So if sisters are going to help each other reproduce, they are surely not going to include their less-related brothers. The very high relatedness among sisters makes forfeiting their own reproduction to help their sisters reproduce pay off, genetically speaking, because their sisters bear so many of the same genes. Multiple fathers decrease the strength of this selection, but once sociality evolves, even this confers some advantages. In any case, the high relatedness among females produces a strong selective pressure to evolve sociality, and this is one reason all-female sociality evolved seven to nine times in this evolutionary lineage.

ANT NESTS

Ant nest architecture originated about 100 to 140 million years ago when the ancestor of all modern ants dug the ancestor of all modern ant nests. Today, the many thousands of ant species and their nests are the descendants of this ancestral ant and her nest. However, in the humid tropics, about half of the ant species nest in trees, often reaching a huge abundance. I once fogged two rain forest trees in Guyana with a "knock-down" insecticide to collect the arthropods living in them. About 90% of the resulting rain of insects onto our collecting sheets were ants, and most were a single, superdominant species of *Azteca*. With increasing latitude both north and south, ant species nest less frequently in trees and more frequently in the ground. This shift probably occurs because of the seasonally harsh desiccating and/or freezing conditions of life in trees in temperate climates. At my latitude in Florida, only a few of the hundred or so species nest only in trees. Most nest in the ground or in rotting wood or make temporary nests in existing shelters.

Each of the ground-nesting ant species builds a more or less distinctive nest, differing not only in size but also in architectural details from those of other species. In modern biology, these differences are best explained by evolution.

A 1973 essay by the evolutionary biologist Theodosius Dobzhansky of fruit fly genetics fame was titled "Nothing in Biology Makes Sense Except in the Light of Evolution." By this he meant that there is overwhelming evidence that all life on Earth is related and can be arranged into family trees that make sense as the outcomes of evolutionary descent. Evolutionary explanations are a core of modern biology, from molecules to behaviors to societies. The diversity of ant nest architecture, like the diversity of organisms, must also be the outcome of evolution and can therefore be arranged (at least in theory) into family trees that show lines of descent of closely related organisms or architectures as branches. To work toward this goal with respect to nest architecture, the reader will have to allow me a good deal of latitude because my sample size is small. Nevertheless, the exercise can be educational and is undertaken in chapter 9. From the outset, I must make clear that what evolved is not the ant nest—which is just hollow space in dirt—but the behavior of the ants that dig the nest. The nest is essentially the product, or "fossil," of the ants' behavior. Much complexity is hidden in this simple claim, for "behavior" includes not only the lone actions of individual workers, but also how these workers are affected by the behavior of other workers and by the cues and feedbacks emanating from the nest as it is constructed by dozens to millions of workers in the dark, without a leader or a blueprint. It is, in the currently popular phrase, self-organizing. How this self-organization works during nest excavation is largely still a mystery but is one of the central questions of the field.

THE MEDIUM

To those of us who move freely in sunlight and air, soil is the dense, granular medium on which we walk, in which plants are rooted, on which our houses are built, and on which we place asphalt in parking lots. We have no experience that would allow us to imagine moving through soil. Most creatures that live in soil move through it by creating cavities, and much of this activity involves excavating soil and dumping it on the surface, out of the way. This is what ants do when they create their subterranean nests. But most soils also have abundant empty pore space, and some soil-dwelling creatures are so small that they can move in the spaces between soil clumps or grains, or they can exploit the ample pore space by pressing grains aside to gain passage. Some ants such as the minuscule thief ants are so small that they are probably capable of doing this. Their wide-ranging, dispersed, threadlike passages are rarely accompanied by soil dumps on the surface. When these passages intersect with the nests of larger ants, the thief ants steal and eat the brood of the larger ants.

In the coastal plains of Florida, soils barely qualify as "soils" among soil scientists, who describe them as mere sedimentary deposits because they show little or no formation of zones or horizons. Indeed, most of the coastal plains are stranded coastal dunes running more or less parallel to the current shore. The lower interdune valleys form a network of sluggish streams that gradually drain the shallow water table to the Gulf of Mexico. An elevation difference of 1 to 2 m separates the wetlands from the dry uplands. The wetland soils are dark, organic, and mucky, while the upland soils are almost pure sand, with charcoal dust darkening the top 10 cm to gray. This soil is so sterile that no farmer, native or immigrant, was ever foolish enough to try to farm it, reprieving the pine forest from destruction. Competition for nutrients is intense because 85% of the nutrients are in the top 15 cm of soil. Getting a share of these nutrients on their way down to the water table requires plants to capture them before their neighbors do. Roots are sparse below about 20 cm.

This is the environment in which I began my field studies of ants, lured by the diverse and abundant ant fauna that proclaimed its presence, like name tags, through distinct patterns of soil dumps. It seemed likely that this surface distinction also reflected distinction underground, and it was this underground distinction that I wanted to reveal.

CHAPTER 2

Revealing the Shape of Empty Spaces

Ant nests are hollow spaces within soil—air surrounded by soil. The notion of filling a hidden empty space to show what "shape" it is must have occurred to people many times through human history. Of course, it works only if the mold can be removed from around the filled space. It first occurred to me that this could be applied to ant nests when I was a young assistant professor in the 1970s. I remembered that the circulatory systems of frogs and sharks that we dissected in anatomy class came filled with red or blue latex, so I ordered some latex and had my undergraduate student Grayson "Chip" Brown try to make a nest cast with it. He poured the latex into the opening of an ant nest in the backyard of his rented house, waited for it to harden, and then dug it out. The result was a really nice cast, like pancakes on a stick, almost 2 m deep, with the problematic feature that it was floppy, made of rubber as it was. We identified the ant as the cone ant, *Conomyrma flavopecta* (now *Dorymyrmex bureni*), and put the cast in a laboratory drawer where it remained for 15 years or more, to be hauled out occasionally for visitors, and eventually disappearing during a move of my laboratory.

My urge to cast ant nests was reawakened in the mid-1980s after two USDA biologists, David Williams and Cliff Lofgren, published a brief paper on the use of dental plaster to make ant nest casts. This orthodontic plaster was much stronger than plaster of paris, dried hard and stiff, and was a huge improvement over floppy latex. Moreover, it was cheap, available in 25- or 50-pound boxes, and mixed with water it set in about 30 minutes. I bought some plaster to make a cast of a fire ant nest. After having excavated over 100 such nests, I expected that the nest would consist of randomly located and connected small chambers. Using a pretty low-tech process, I poured a half bucket of plaster slurry into a fire ant nest, dug up a 20 kg hunk of dirt and plaster, and took it home to wash off the soil with a hose to reveal the shape of the empty space. It was an eyebrow raiser, for I hadn't imagined it correctly at all. In place of the rather random structure I had imagined, the cast revealed a clearly organized structure of repeated vertical shafts connecting horizontal chambers into many distinct vertical series, often with horizontal connections between chambers as well (fig. 2.1). I hung the cast from the ceiling of my office, where it amazed visitors for three decades.

This minor epiphany opened my eyes to the potential of effective nest-casting methods for studying ant nest architecture. This potential was appar-

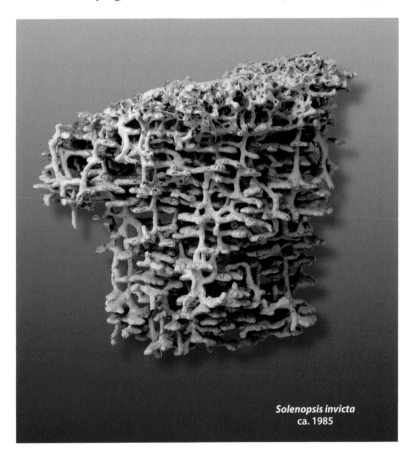

Solenopsis invicta
ca. 1985

Fɪɢ. 2.1. The first plaster cast of a fire ant (*Solenopsis invicta*) nest. Even though it is incomplete, the obvious organization revealed by this cast opened my eyes to the value of making casts of nests, rather than simply excavating them. Author's photo.

ent even from the single plaster cast I had made—in three dimensions, it revealed every detail of architecture the way a print reveals every detail of a negative. Up until then, the occasional studies on ant nest architecture had relied on sketches or scale drawings of excavations (fig. 1.2), but imagining these nests in three dimensions was challenging. In comparison, viewing a three-dimensional nest cast seemed almost magical.[1] It could be viewed from any angle, with each view contributing to an image of the whole. Plaster certainly had its limitations, such as fragility, but for the time being, plaster set me on the road to studying ant nest architecture. Along the way, this basic method, combined with other methods, allowed me to delve into related areas of ant behavior, colony biology, evolution, and ecology, all of which I will touch on to paint a more unified picture.

THE CAST THAT GOT ME HOOKED

In the next few years, I sought out and cast a handful of nests of several ant species, but my hands were full with my fire ant research, so this stayed an interesting sideline until, in 1999, I poured a five-gallon bucket of dental plaster slurry into a recently abandoned Florida harvester ant nest. It took the full five gallons, filling the nest to a depth of 2.5 m. Digging it out took two days and resulted in almost 150 pieces of cast, a giant, three-dimensional jigsaw puzzle that I carried home in several large trays. In retrospect, my decision to mount and reconstruct this cast was founded on my complete lack of imagination regarding what it would take to put all the pieces together. I built a plywood base and back, painted it blue, got a lot of thin, stiff metal rods, found a location for the cast in the building in which I worked, and started the reconstruction. The cast had to be constructed from the top down, so the first problem was to estimate how tall it would end up being. I settled for just short of eight feet, as that was the length of the plywood sheet and the depth of the

1 During the 1950s, Meinhard Jacoby made huge nest casts of Brazilian leafcutter ant nests using cement, but even these remarkable casts were published as drawings rather than photographs.

hole left after the excavation. In my excess of enthusiasm during excavation, I had failed to mark the origin of the many pieces, and this was further complicated by the presence of four separate shafts. Just as in assembling blue sky in a jigsaw puzzle one looks for small variants in the outlines of the pieces, the precise shapes of the breaks could be fit together without doubt after trying many combinations. Another feature that sorted chambers into distinct and mutually exclusive groups was the color they took on from their location in the soil column. Like the soil in the colony's sandhill site, the chambers from the top 15 to 20 cm were dirty gray, those from the next meter or more were yellow, and then they became pure white. So I never had to test whether a white piece fit with a yellow or gray one. It also soon became apparent that the shafts were smooth helices, like spiral staircases, so junctions that made sharp changes of direction were clearly incorrect. By placing the broken ends of candidate pieces together and gradually turning them, I would suddenly make a perfect fit. The only thing missing was an audible click. Then came "the mixing of the epoxy," followed by gluing the pieces together, with careful support to maintain the exact position until the glue hardened.

The top region of the nest was so interconnected that it was recovered as one single, if very complex, piece. I placed it upside down on the lab bench and reconstructed the cast upside down for about 30 cm. Things were going well until my student lab assistant spun around with a backpack and shattered the partially reconstructed cast into about 50 pieces. Like Sisyphus, I started up that hill again, and after a couple of weeks, step by step, the cast regained its earlier level of reconstruction, though with more glue joints.

At this point it was necessary to mount the cast against the backboard by means of stout metal rods, a tense process during which my daughter Erika helped hold and stabilize it. With the upper cast region in place, reconstruction continued downward, each piece supported by a metal rod projecting from the backboard, each glue joint having to fully harden before the next piece could be fit. I also built some subassemblies separate from the main cast and added them as more complex units. In this way, the cast gradually took form over two months or more. It was not work for the fidgety.

Fig. 2.2. The first complete plaster cast of a Florida harvester ant nest. This is still one of the largest casts I have made. Photo by Charles F. Badland.

The result, however, was magnificent, revealing an unsuspected beauty and order (fig. 2.2) that made me want to reveal more such beauty in other nests and display it for others to see. There were two large barriers to realizing this desire—first, the daunting amount of work it took to reassemble a large plaster cast, and second, the obvious fragility of the reassembled cast. There was no way such a cast could be moved.

THE SEARCH FOR BETTER MATERIALS,
AND SOME ADVENTURES

Thus began the hunt for a stronger casting material. Adding glass fibers or other additives to the plaster helped a little, but not enough. Hardening plastics were out because they were true liquids and would soak into the sand and make an unnaturally bloated cast. It had to be metal, but which metal, and how to melt it in the field, right next to the nest to be cast? Having stayed awake in high school physics class now paid dividends—I recalled Newton's law of cooling, which said that the rate of heat transfer increases with the temperature *difference* between a hot and a cold material. Any molten metal poured into a nest would lose heat to the surrounding sand as it flowed down, and other things being equal, the higher the melting point, the faster the rate of heat loss and the sooner the metal would freeze. Therefore, high-melting-point metals would fill much less of the nest before freezing than low-melting-point ones.

Reviewing the physical properties of metals suggested only two realistic candidates with sufficiently low melting points—aluminum and zinc. Lead melted at a lower temperature, but its softness and toxicity removed it from the list. Aluminum was obviously easily available as scrap, but what about zinc? Where does one get zinc? On the world market, it was a fairly expensive metal and beyond my budget. At this point, an interesting factoid from college chemistry provided the critical piece of information—zinc ingots connected to the hulls of steel ships prevent hull corrosion in seawater by corroding the zinc, the so-called sacrificial anode. I called around to boatyards in the area, and yes, they put little blobs of zinc on outboard motors, but there wasn't enough, nor was it available. Consulting the yellow pages, I found the Atlantic Marine Shipyard north of Jacksonville, Florida, and spoke to an engineer named Leon who said yes, they had sacrificial zinc anodes they were recycling, and yes, I could have some.

I wasn't prepared for the scale of the shipyard and what they did there other than replacing half-used sacrificial anodes. They repaired and modified aircraft carriers, dry-docked a range of ships, and even built ships from the ground

FIG. 2.3. *Left*, zinc anodes, whole and broken into usable pieces. The small pieces weigh roughly 0.5 to 2 kg. *Right*, a discarded aluminum scuba tank and a tank cut into pieces ready to melt in the crucible. Author's photo.

up, so to speak. An almost finished ocean-going tug was sitting on blocks near the water. A huge fill-dumping ship with a hull that was hinged lengthwise on the deck so the ship could be opened like a clamshell was almost ready to slide back into the sea. In a quarter-mile-long building, computer-controlled plasma torches were cutting 3 cm thick steel plate at over a meter per minute, while automated welding machines began assembling the pieces into ship parts. Behind the office building were two large dumpsters filled with zinc anodes, each coated in a thick white layer of corrosion (zinc carbonate) (fig. 2.3). "Help yourself," said Leon. "We usually sell to a recycler, but you can have what you need for free." A fabulous shipyard tour plus free zinc—what could be better?

Zinc is a dense metal—a cube 22 cm on a side weighs 70 kg, about the average weight of a human, so by the time I finished loading it, my poor little Toyota Corolla wagon looked like it was waddling, but I ignored its cries of pain and drove back to Tallahassee, though at reduced speed. My zinc supply was now secure, but I also wanted to try to melt aluminum because it was lighter, stronger, and more available. I acquired a modest pile of scrap aluminum from Leon Scrap Metals and soon learned that there were hundreds of alloys of aluminum, all mysteriously differing in their properties. Adding less than 1% of silicon or iron to pure aluminum doubles its breaking strength and increases its stiffness by over tenfold. Adding a small percentage of various other metals like magnesium, manganese, and copper increases the breaking strength by tenfold to fifteenfold, and stiffness by up to a couple of hundredfold. In the end, the only thing that mattered in these early trials was that I could melt the metal easily.

Now I faced the problem of how to melt it in the field. Knowing that even today, traditional iron making in Africa is done in earthen furnaces fired with wood, I first tried it with a pit dug in the sand, a Dutch oven, and a bag of charcoal briquettes. It was a resounding flop—although I turned some sand into lumps of low-grade glass, not the first drop of liquid aluminum appeared. It looked like I needed to restrict heat loss by moving the fire into a more enclosed cavity so that less heat would escape and the cavity would approach thermal equilibrium. The temperature of the cavity would then be revealed by the color of the emitted light, changing from infrared to dull red, red, orange, yellow, and white as the temperature increased. A small booklet purchased on the internet provided plans for a gas-fired foundry built from a 5-gallon paint drum. I bought a 20-gallon garbage can and lined it with a mixture of sand and fire clay, insulating the kiln against rapid heat loss. It took a week to dry and weighed over 140 kg, so I couldn't easily heave it around by myself. Inside this insulated cavity, I added a cage that created a space for charcoal between the wall and the central cavity, into which I slid the crucible made from the bottom half of a steel scuba tank provided with a bucket handle. The purpose of this cage was to prevent the charcoal from collapsing into a heap when the crucible was removed, making return of the crucible difficult. Under this was an ash grate with openings below to allow air to enter below the fire.

The first version of this weighty kiln called for a draft of air into the bottom, something like a blast furnace, so I made a trip to the You-Pull-'Em junkyard with socket wrench and screwdriver in hand and removed the heater fan from a 1989 Chevy. When I hooked this up to a 12-volt marine deep cycle battery, the fan blasted a hurricane into the bottom of the kiln, causing the burning charcoal to illuminate like light bulbs. During the first test run, aluminum started puddling in the bottom of the crucible within about 45 minutes, and in an hour, I was ready to pour.

This success meant I had to haul around a good deal of equipment. In addition to the 140 kg kiln, there was the battery, the blower, a heavy kiln lid of refractory bricks, 20-pound bags of charcoal, scrap aluminum, Kevlar hot gloves, and diverse tools for stirring, deslagging, lifting, and pouring. I usually borrowed the departmental pickup truck for moving all this stuff, and of course I needed another strong back to heft the kiln (although my friend and colleague Josh, a national shot put champion, could heft it by himself). Somewhere along the way, I discovered a marvelous refractory blanket material online that was capable of withstanding temperatures of 1200°C. So I took a hammer, knocked out the sand–fire clay insulation, and replaced it with 10 or 15 cm of this Durablanket. Now I could easily, although awkwardly, carry the 15 kg kiln for some distance. It still required a heavy battery and a fan (fig. 2.4).

Fig. 2.4. The semifinal version of a kiln in which air is provided by a fan powered by a 12-volt battery. Author's photo, from Tschinkel (2010).

Fig. 2.5. Three sizes of kilns, each shown with a typical size of crucible. The left kiln is fitted with two computer fans to provide a draft. A motorcycle battery provides the 12-volt power. In the two right kilns, a draft is created with a chimney. Air is drawn through vents under the ash grate near ground level. Author's photo.

The next improvement was an alternative and less equipment-intensive way of providing air to the fire. On the theory that a draft could be generated either by pushing air (a blower) or by pulling air with a chimney, I cut vents near ground level and added a 2 m tall stovepipe chimney to the lid. The reduced pressure created by the expanding hot gases escaping up the chimney drew in air at the bottom with a reassuring whoosh. It turned out to be just as effective as the heavy battery-fan combination, weighed very little, and never required recharging (fig. 2.5, *right*). The kiln produced three to five liters of glowing aluminum in one to two hours.

Yet another improvement was the use of boron nitride paint to protect the inside of the steel crucible from the strong corrosive effects of molten metal, especially zinc. It works because boron nitride is not wetted by molten metal, preventing contact between the melt and the crucible wall. A problem that still plagues the process even today is the burning of the outside of the steel crucible in the high heat and oxygen-rich conditions in the kiln, causing burned metal to flake off and thinning the crucible walls. An anticorrosion paint has slowed this burning, but it remains a problem. Because molten zinc is a good solvent of steel, the combination of solution on the inside and burning on the outside created holes in more than one crucible, spilling molten zinc into the bottom of the kiln and making a mess. My friends Sandy and Ralph in the biology machine shop got tired of cutting the bottoms off steel scuba tanks and providing them with bucket handles to replace the dead crucibles.

Much scarier was the time we once filled the crucible with enough zinc to contact the steel handle, weakening it so that when we lifted the crucible out, the handle failed, dropping the crucible with 35 kg of molten zinc back into the kiln and splashing zinc 2 m to one side (but missing both of us). My technician and former student Kevin Haight and I both gulped, retrieved the crucible, and finished the pour. We never overfilled a crucible again.

The zinc supply problem had been solved with two visits to the Atlantic Marine Shipyard, but collecting enough scrap aluminum was challenging. Knowing that scuba tanks were made of either steel or aluminum and that both types had to be tested regularly, I scoured dive shops to find aluminum tanks that had failed the test and needed to be scrapped. Sandy and Ralph in the biology machine shop band-sawed these tanks into pieces that would fit into my crucible (fig. 2.3, *right*). At 15 kg per tank, this was a bountiful supply to melt and cast. Business was good! At the time of this writing, perhaps 20 or more scuba tanks have ended their life as a pool of red-hot aluminum in the bottom of my crucible before being transformed into a beautiful cast of an ant nest. A noble fate, in my opinion.

I learned the hard way that kilns need to be watched. My friend Dennis and I had just poured molten aluminum into a nest, put the half-full crucible back into the kiln, and put the lid back on, and we were concentrating, heads down, on digging up the cast. After 15 minutes, I glanced at the kiln about 15 m away and was shocked to see a jet of yellow incandescent gas 2 m tall emanating straight up from the smoke exit in the lid. It was like a jet of ionized gas blasting from a black hole in some distant galaxy, glowing with a brilliant yellow light. We froze for a moment and then rushed to whip the lid off the kiln, discovering that the extreme temperatures had burned a hole in the steel crucible and dumped the rest of the aluminum into the bottom of the kiln. I tipped the kiln to drain the aluminum onto the ground through an air hole.

The biscuit of solidified aluminum sits in the bottom of my kiln to this day, but ever since then I have been careful to pay attention after I put the lid back on. Understandably, these kinds of adventures are best done in remote areas (the "boondocks"), away from populated areas. Fortunately, I live in the American South, where doing strange and dangerous things in the woods is considered normal.

GRADUAL IMPROVEMENTS

Step by step, improvement by improvement, a final version of the kiln took shape. Applying the same principles of construction and function, I built kilns in smaller oil drums, galvanized pails, and even a coffee can in which I melted zinc in a soup ladle. Each was more suitable for certain sizes of ant nests. Most further improvements have been responses to various quirks of the present kiln versions. For example, all suffered from the buildup of ash on the surface of the charcoal, reducing the diffusion of oxygen to the charcoal surface and therefore the rate of primary combustion, a problem that can be solved by simply shaking to dislodge the ash into the ash bin. This maintains the maximum rate of combination of oxygen with the solid carbon surface to produce carbon monoxide, which in turn combines with more oxygen to produce CO_2, lots of it, and heat, also lots of it. Simple chemistry tells us that 12 kg of charcoal will combine with 32 kg of oxygen (carbon *dioxide*!) to produce 44 kg of carbon dioxide.

When I set out on a nest-casting expedition, all the equipment now fits into the back of my Subaru Forester, with an empty front passenger seat for a companion. Equally important, when the cast is done and the glowing remains of the charcoal are buried in the pit from which we took the cast, the low mass of the kiln means that with the lid removed, it cools down fast enough to load into the car in about 20 minutes. The first model kiln, which was insulated with 140 kg of sand–fire clay, was still uncomfortably warm after eight hours. I no longer need to borrow the departmental pickup truck, and casting expeditions have become somewhat routine—choose the appropriate-sized kiln and

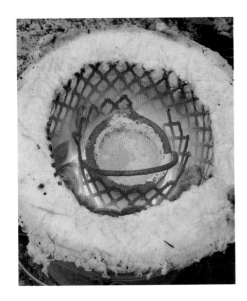

Fig. 2.6. A small graphite crucible of red-hot aluminum ready to pour. Note the fire-resistant kiln lining. Author's photo.

load it into the car along with charcoal, metal, and diverse tools (although I invariably forget some item and must improvise).

Nowadays, I no longer use matches to light the fire—in the humid Florida climate, even strike-anywhere matches soon become strike-nowhere matches—so now I start the fire with a self-igniting propane torch. The kindling in the kiln offers only futile resistance to the fire blast emanating from the torch. When some of the charcoal is aglow, the crucible goes into the cage, the metal into the crucible, and the chimney does its work of firing up the whole charcoal mass into an orange glow so that aluminum begins to melt at dull red heat in about 45 minutes. In another half hour, the crucible and metal glow orange yellow like a 40-watt bulb and are ready to pour (fig. 2.6). Heating the aluminum so far above its melting point ensures that it will stay liquid longer and therefore flow deeper into the nest, as well as penetrate even tiny details of nest structure.

Admittedly, fishing a crucible that is glowing like a light bulb out of the kiln is a bit daunting (fig. 2.7), and my heart always speeds up as the moment approaches. Radiation from the crucible is intense enough to light fire to any dry organic matter within 10 to 20 cm (fig. 2.8). Needless to say, wearing shorts is

FIG. 2.7. Pulling the red-hot crucible out of the kiln. Author's photo.

FIG. 2.8. Pouring molten aluminum into a harvester ant nest. Radiation from the crucible has ignited the surrounding vegetation. Tongs allow the two pourers to put some distance between themselves and the crucible. Author's photo.

FIG. 2.9. Emergency heat shield fashioned from trash on an occasion on which, in a moment of forgetfulness, I wore shorts. Photo by Christina L. Kwapich.

not smart, although I did once accidentally and had to improvise heat shields from some sheets of Tyvek foam insulation board that someone had left behind (fig. 2.9). Wearing Kevlar thermal gloves is necessary, not so much because one is likely to be careless enough to touch a red-hot crucible, but because the radiation at such close quarters would burn bare skin. My crucibles have a little loop at the bottom edge so I can tip and pour from medium and small crucibles by myself. Pouring a large crucible holding five to seven liters of aluminum and weighing 15 kg is a two-person job in which the crucible is clamped between tongs that allow the two pourers to be farther from the crucible (fig. 2.8). During second and third pours, sharing a pit with a glowing crucible runs the danger of lighting your socks on fire, and you begin to understand how the law of radiation—that is, the decrease of intensity with the *square* of the distance—works in your favor. When the pit is very deep, it is simply too dangerous to have someone lower a glowing crucible full of liquid metal down while you are in the pit, and equally dangerous to jump in after such a crucible.

FIG. 2.10. Pouring a third time from the edge of the pit. This is far safer, though less accurate than being in the pit with the crucible. Author's photo.

In response to this dilemma, I use a seven-foot hay hook to lower the crucible, which is endowed with a small metal loop near the bottom so I can tip it with a long, hooked metal rod and safely carry out the whole pouring operation from the edge of the pit (fig. 2.10). My aim is not always good, but at least I still have my socks.

The techniques I developed for metal casting have been copied by a number of people who saw my work on the internet. Some of them are hobbyists who offer nest casts for sale online, usually of fire ants, and show little interest in the biology of their victims. To my knowledge, the first scientist to make a cast of an ant nest for scientific reasons was Meinhard Jacoby, who made cement casts of the fungus-gardening ants *Atta* and *Acromyrmex* in Brazil from the 1930s to 1950s. From this beginning, several researchers in Brazil and Argentina, first and foremost Luiz Forti and his colleagues, have used cement casting followed by excavation to reveal many of the enormous nests of several species of fungus-gardening ants. These monumental casts may require up to 10 tons of cement and take a couple of months to dig up. Only a totally numb person would be unimpressed by the scale of these nests (and casts).

Interestingly, two cousins in Australia, Christopher East (an industrial chemist) and Stephen East (a builder), have also persisted in making casts of a number of identified ant species (Australian.ant.art.com). Although their initial motivation was not primarily biology, they have contributed to an inventory of Australian ant nest architectures. Several Australian ants are very large and have correspondingly large nests, requiring several hundred kilograms of molten aluminum and a can-do attitude to cast.

In addition to these practitioners of nest casting, there is a sizable group of researchers whose primary interest in nest architecture is theoretical, based on modeling and laboratory studies. Because my research in ant nest architecture addresses what the ants do in nature, the work of these theory and laboratory researchers is outside the scope of this book. I have included a few key citations in the references section.

Having refined my casting and digging technique, I now had the ability to make the invisible visible, to make perfect solid objects out of empty space. But who were the creatures that created these hollow spaces? How did they occupy the spaces they made? In the next chapter, I will take the reader through an excavation of a live harvester ant nest to reveal the architects of these beautiful nests.

CHAPTER 3

Meet the Architects

Making an aluminum, plaster, or zinc cast renders the architecture of the nest beautifully, but it tells us little or nothing about where in the nest ants can be found. The biological world consists of structures from molecules to ecosystems, so there is good reason to assume that ants are structured and organized within the nest rather than distributed willy-nilly. What if workers assort themselves into specific zones or chambers of the nest where specific tasks are carried out? What if such structure is important to how the colony works, to how it carries out life functions efficiently? What if the life cycles of both the colony and the individual ants are somehow linked to organization within the nest space? The only way to find out is to dig up ant nests and describe what we encounter.

I started digging up nests of harvester ants (*Pogonomyrmex badius*) in the 1980s and eventually developed a consistent procedure for collecting the ants and nest contents chamber by chamber. A number of small adjustments and improvements made the process more effective and faster, so I could be confident that my efforts and those of my students recovered everything that was in the nest—ants, seeds, and assorted guests.

DIGGING A LIVE HARVESTER ANT NEST

The procedure of digging up harvester ant nests is well illustrated in a dig from 2016. I loaded shovels, two sizes of brick trowels, lots of plastic boxes, trays, and bags, a couple of aspirators, a large tarp, and a battery-powered shop vacuum into the back of my Subaru Forester. A big blue canopy with aluminum legs unfolded into a shade tent for working.

From my laboratory in Tallahassee, Florida, I headed southwest for seven miles on Springhill Road and then west into the Apalachicola National Forest to what I refer to as Ant Heaven. You won't find this name on any map—I named it for its abundant and diverse ant population—but you will find it in my published papers. The site was largely clear-cut in the 1970s and then partly replanted with longleaf pine and partly left to itself. Lying between the flatwoods to the west and the deep channel of Fisher Creek to the east, this sandy site is excessively drained, making it congenial to drought-loving animals such as harvester ants and gopher tortoises, and plants such as prickly pear cactus and beargrass. A small, easily missed track off the forest road leads to Ant Heaven. Sometimes the overflow from the cypress swamp is deep enough to come up to the car doors, but on this day in 2016 it was dry, as there had been little rain for a couple of months. I parked in the usual place near a patch of sandhill milkweed. Nearby, the shiny blueberries were ripe, so I could take breaks for some sweet snacks. On the drive in, I was relieved not to see any new litter of Bud Light cans, shotgun shells, shot-up TV sets, or broken whiskey bottles. We periodically clean such litter up because it attracts more litter.

Finding the nests of harvester ant colonies might be hard on a moonless night without a light, but in daytime, it is like falling off a log. This is because the nest disks are not only large but are covered with bits of black charcoal with which the ants decorate their nests for unknown reasons (reasons that will remain unknown for the foreseeable future, as we have tested eight different hypotheses and found none supported). I picked a nest not too far from the parking place so I wouldn't have to carry my stuff too far.

The nest was actively foraging, and some of the foragers were away from the nest. Throughout my dig, I grabbed them as they returned so the final nest population count would be correct. If I had gotten started before 8:00 a.m., I would have found the nest closed and all the foragers still inside, and I wouldn't have had to grab them as they returned from foraging. As a first step, I scraped off the charcoal and captured all the ants on the surface with the shop vacuum, a gadget whose true value you will understand as the dig progresses. Now I unfolded the tarp, laid it out to one side of the nest, and started digging a pit

FIG. 3.1. An early stage in the excavation of a harvester ant nest. Excavation proceeds from a pit on one side of the nest. Ants are collected with a battery-powered vacuum, which also aids in exposing and cleaning chambers. Frame from a video by Henry M. Tschinkel.

about 50 cm from the nest (fig. 3.1). The pit needed to be a lot larger than you would think, for when it is deep, it must be possible to swing a short shovel within it. Avoiding the work at the beginning exacts a cost later when one has to enlarge it when it is 2 m deep. This is advice that I and my helpers have often ignored, with predictable (and avoidable) consequences.

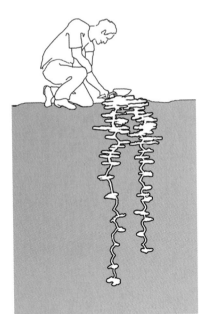

When the pit was about 50 cm deep, I squared up the sides and leveled the floor. In my opinion, a neat pit is a sign of a neat mind. I developed this opinion in the late 1980s when I and my assistant Natalie Furman excavated almost 40 colonies in a year. Archaeologists make very neat, clean pits that allow them to keep detailed track of what they find, and I think of the dig as myrmecological archaeology as I dig up and collect artifacts, noting their location and nature. It is important to toss all the excavated dirt some distance away onto the tarp, because when the pit is deep the piles will be high and the sand will slide back down on one's head, and because one will be sweaty, it will stick all over.

Next, I carefully cut back the wall to the nest until I just nicked a nest chamber and a few workers poured out. This usually happens just under or just outside the edge of the charcoal layer. I plugged the break with sand, and using a large trowel, I began to lift off the thin layer of soil that covered the very complex top chamber, working carefully toward the edges, where fingerlike extensions housing pupae often occur. Now, as I sucked up the panicked workers and pupae, it became even clearer what a great tool the shop vacuum is. When I first started excavating harvester ant nests, we used a mouth aspirator to collect the ants— my teeth were crunchy for hours afterward, and my spit was black and sandy. I dumped the captured ants and brood into the tray labeled "surface to 10 cm." A few ants were a lighter reddish brown among the hundreds of darker workers. As I dug deeper, the frequency of lighter workers increased.

The vacuum is also excellent for cleaning fallen sand from the smooth floors of the exposed chambers, which is very hard to do otherwise. Careful vacuuming can reveal the chamber outlines, ready to be traced or photographed. For tracing, I placed a sheet of transparent acetate on top of the exposed, cleaned-up chamber and drew its outlines with a marking pen (fig. 3.2). I noted the depth of the chamber from the surface, and its horizontal location in reference to a zero-zero point. This way, I had its three-dimensional Cartesian coordinates, x, y, and z. I did this with every chamber I exposed until I hit the very last chamber, which was a meter or two below my feet. I also put the

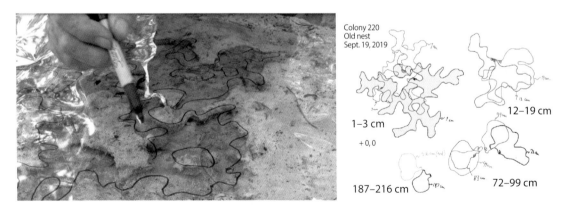

FIG. 3.2. Tracing the outlines of exposed chambers on transparent acetate. Four examples from different depths are shown at the right. Frame from a video by Henry M. Tschinkel.

Fig. 3.3. The exposed uppermost chamber of a *Pogonomyrmex badius* nest. In this image, the chamber floor has been digitally darkened to increase its contrast with the surrounding soil. Frame from a video by Henry M. Tschinkel.

contents of each chamber into a separate container, labeled with the depth. As the dig progressed, I continued to vacuum up every worker I saw, especially those that fell into the pit with me, for their stings are quite painful.

Having exposed and cleaned up the topmost chamber, I could appreciate its interlacing complexity, as though composed of branching tunnels widened to form large fused spaces (fig. 3.3). Here and there were a few seeds and pieces of insects recently dropped off by foragers. My manipulations knocked a couple of alleculid beetle larvae from their burrows, and they desperately humped along trying to find a place to hide (fig. 3.4). Before I messed with the nest, they were ensconced in little burrows in the floor of the chamber, waiting to steal insect prey brought in by foragers. It is easy to

9.5 mm long

Fig. 3.4. *A*, alleculid beetle larva, *Hymenophorus tschinkeli*; *B–D*, mites; *E*, silverfish; *F*, a collembolan; *G*, a *Masoncus pogonophilus* spider. *A–F*, Author's photos; *G*, photo by Paula Cushing, with permission.

show that this is their game by giving the foragers blue-dyed insect prey to take back to the nest. An hour later, you can collect blue alleculid larvae from the upper chambers.

These top chambers are hard to assign to a single depth because they warp and blend, but one can also see several distinct openings that connect levels. Some of these connect only one or two levels, so I had to figure out which one (or ones) descended deeper into the nest to form the main shaft or shafts connecting long series of chambers, some all the way to the bottom. Using the large brick trowel with sharpened edges, I carefully lifted off the soil to expose the next level of chambers below. I collected the panicked ants and cleaned up the fallen sand with the shop vacuum. The scent of sour apples rose from the scurrying ants, a scent that wafted up repeatedly during the dig. This is the odor of the alarm pheromone (4-methyl-3-heptanone) the ants release when they are disturbed, and it mobilizes their nestmates to rush around to discover and attack whatever caused the disturbance. Its effects are short-lived, as it soon evaporates to nondetectability.

The chamber I had just exposed was a little simpler and smaller and had only three openings descending to deeper chambers. I made an acetate tracing of the chamber and marked the entrances to deeper levels. So as not to waste acetate, I traced four chambers on each sheet by using different colors. In these chambers were patches of clean yellow sand within the generally gray sand at this level. These were chambers the ants had backfilled with yellow sand from deeper layers. They do a lot of this in the top several chambers, as we will see in chapter 7.

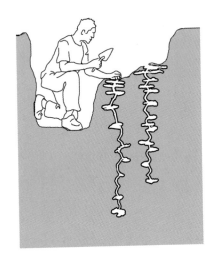

It seemed like slow downward progress because the chambers were so close together—at only about 10 or 12 cm belowground, I had already exposed and traced four or five levels. But now the chambers were down to only two entrances to continuing vertical shafts, so I got busy with the trowel and exposed the next chambers.

The chamber at 16 cm was smaller and less fancy than those above (fig. 3.5), and it contained a half dozen seeds that were probably being transported to lower levels when I took the roof off the chamber. One was a beautiful, shiny round seed that was so large it looked as though the ants couldn't possibly grip it to carry it. Oddly, the ants seemed very fond of these catbrier seeds, even though given the maximum gape of their mandibles and the size of these

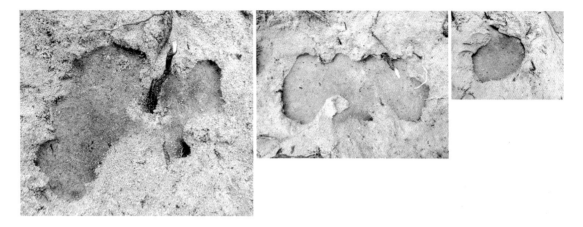

FIG. 3.5. Chambers become smaller and less complex with depth. The darkened chamber floors are probably due to a fungus. Author's photo.

smooth seeds, there was no way they could crack the seed coat to get at the edible seed inside. There was also a giant horse nettle seed.

I lifted off another 8 cm of sand, tracking the shaft downward, and noticed that a patch of sand had sagged slightly because there was a chamber just below. The utility of the vacuum now became apparent, for I simply sucked this sand roof off, exposing the clear floor of the chamber. A few seeds lay scattered about, and a silverfish scurried in panic looking for a place to hide. I used the vacuum to remove fallen roof material all the way to the edges of the chamber. A few workers appeared out of the shaft to the next level, and I sucked them up.

You may be thinking that my efforts to characterize the vertical distribution of workers in the nest were pointless because the excavation was driving the workers downward to escape the catastrophe approaching from above. When I started digging up harvester ant nests in the late 1980s, this possibility made me queasy, so I tested to see whether it had any merit. I dug a 2 m pit close enough to two colonies to just nick a chamber or two. After plugging the nicks, I hammered in pieces of sheet metal at several horizontal levels to sever the shafts connecting the chambers, trapping the ants between these barriers. The next day, I excavated the nest and discovered that the vertical distribution of dark and light workers, seeds, and brood was not different from that in nests without such barriers. Moreover, if workers were really trying to move up or down as we dug, they should have accumulated above or below the barriers, and they did not. So, strange as it may seem, the arrangement I observed during my dig is a pretty good representation of the undisturbed arrangement.

There were fewer workers as I revealed deeper and deeper chambers, but I knew there would be plenty more as I approached the bottom of the nest. At 45 cm I hit the first seed chamber, and an impressive one it was. I cleared it to the edges to reveal a large kidney-shaped chamber almost 15 cm long. The floor was almost completely covered with several layers of seeds of many sizes and species (fig. 3.6). The seed layer was so thick that when the ants walked on top of it, they must have bumped into the ceiling.

I bent down to look at the seeds close up and saw an army of little white springtails, tiny wingless insects (Collembola), hopping around (fig. 3.4). Every nest has a burgeoning population of them, and it's likely they eat fungus that grows on the seeds. I also saw a tiny spider scurrying to safety. This was

FIG. 3.6. A seed chamber nearly filled to capacity. Note the shaft to the next deeper chamber at the upper margin of the chamber. Author's photo.

Masoncus pogonophilus, a predator of springtails (fig. 3.4) that spends its entire life in *Pogonomyrmex badius* nests, even following the emigration trail to the new nest when the colony relocates. When it comes time to reproduce, the mother spider packs her eggs into a little indentation in the ceiling and spins a cover over them to make the location seem like just more flat ceiling. Because they have such tight bonds to their home colonies, the spiders from neighboring nests are genetically more similar to each other than to those from distant nests, and they probably usually disperse only small distances within neighborhoods.

I scraped together the seeds and bagged them for later counting, sorting, and weighing in the lab. Here at Ant Heaven, harvester ants collect the seeds of over 50 different plant species, many too large for them to open. The large decorative flat ones are prickly pear, and the abundant black egg-shaped seeds are three-seeded mercury. Whereas the ants can crack open small grass seeds, most of the seeds in this and other chambers were too large for them to open. The ants can eat these large seeds only once they germinate within the storage chambers. The challenge is that wild seeds usually don't germinate all at once but spread their germination over multiple years as a hedge against bad conditions. Seeds thus germinate at rates that vary greatly with species and season. The huge stores of large seeds improve the chances of a steady supply of edible germinating seeds, clearly a long-term gamble.

The next chamber was at 65 cm and was also filled with seeds, followed by an empty chamber and then a smaller one at 85 cm that was chockablock full of seeds. A few workers wandered around uncertainly on the blanket of seeds while the usual army of springtails bounced all around them and another spider scampered for cover.

At this point I needed to deepen the pit, mainly to bring the nest to about waist level where it was more comfortable to work. With the pit over 1 m deep, the importance of a wide pit became apparent as I needed space to swing my short-handled shovel. Hitting the pit edge during a swing would shower me with sand.

The chambers were now much farther apart, so I was getting deeper faster. This region was also relatively less occupied by ants, and the chambers at 90

and 94 cm had only three ants between the two of them. I was now tracing a single shaft downward between these smaller and simpler bean-shaped chambers. It is not so obvious in a dig, but the shaft always turns in one direction, sometimes clockwise, sometimes counterclockwise, depending on the colony and region. This helix is visible in the cast of the harvester ant nest (fig. 2.2). I slipped a pine needle down the shaft to make tracing it easier.

Time to deepen the pit again, this time to 150 cm. Pitching the sand out was now more challenging, and it was obvious why I started the piles so far from the pit. My chin was now even with the ground, and the next chamber was about the level of my thighs, at 100 cm. It did not take long to get down to a chamber at 120 cm, and at this point I needed to deepen again, this time going for broke to 200 cm. Until now, the sand had been a yellow buffy color, but below about 150 cm, it became pure white. The yellow is not the color of the quartz sand, but a coating of limonite (ferrous oxide) on the grain surfaces. Their uncoated color is the white I now saw below 150 cm.

Now the ground level was over my head, but the chambers were again at a comfortable working level. However, I was now gambling with having a wall collapse, burying me up to the knees and obliterating my newly exposed chamber. Collapses are hard to predict but fortunately are not common. At this depth, the importance of the short handle on my shovel now came into focus because it allowed me to swing the shovel inside the pit. My affection for this shovel has increased since I first acquired it from a barn where it had resided since the late 1800s. In contrast to modern shovels, it had a welded (not pressed) shank and a thin, wide blade. It also had a knot in the middle of the handle, so it eventually broke, as though it knew I would need a short handle to work in a pit. I nicknamed it Excalibur, the World's Greatest Shovel (fig. 3.7). The thin blade allowed me to sharpen it to kitchen-knife sharpness, making short work of roots, but its thinness also eventually caused it to develop a split at the edge. To prevent the split from elongating,

FIG. 3.7. *Left*, Excalibur, the World's Greatest Shovel. The blade is short because it developed a split, which I repaired by sawing off the split edge and resharpening it. *Right*, Excalibur in action. Note the shovelful of sand in midair. Author's photos.

I sawed the blade a bit shorter and resharpened it. I did this two more times as the shovel and I both aged from 1975 to 2019, so that the reduced blade capacity paralleled my declining strength in heaving large amounts of dirt out of the pit (fig. 3.7). Over the years, I estimate that this shovel and I have moved around 1,000 metric tons of sand out of and back into pits. My bond with this shovel is strong, but I couldn't say whether Excalibur reciprocates that affection.

Now at 120 cm, I had just removed the ceiling from the first brood chamber, and workers were scurrying around in a panic to carry the larvae and pupae to safety (fig. 3.8). Also, a substantial fraction of workers were now lighter brown. These were so-called callows, younger workers whose cuticle had not fully pigmented yet. In summer-born workers this darkening takes a few weeks, but in autumn-born workers it takes months, and many are still noticeably lighter in the spring. These callow workers carry out much of the brood care, as we will see in chapter 8.

FIG. 3.8. *Upper left*, a newly exposed brood chamber. *Main image*, a detail of workers rescuing brood. Many of these workers are pale young workers called callows. Author's photos.

The chambers were now mostly 10 to 20 cm apart, and I was getting deeper faster. At 180 cm I deepened the pit again, noting that it now took a special movement with my short shovel to make the sand fly out of the pit and far enough from the edge that it didn't slide back on my head. If I hit the pit wall during this pitch, I would be covered in sand. I also needed to widen the pit because the unintended downward tapering had decreased my foot room to an awkward size. The pit was now 240 cm deep.

The chamber at 190 cm had a few larvae and pupae along with a mixture of dark and callow workers. The brood were mostly along the chamber edges, and workers hustled to rescue them from the sudden light. These workers and brood had never experienced light in their lives, and they were programmed to seek darkness. Only toward the end of their lives as they became foragers would they be drawn toward the light.

Finally, I was starting to see a lot of brood. The chambers at 210, 220, and 230 cm were packed with hundreds of workers, many of them callows, and the floor was largely covered with brood (fig. 3.9, *left*), including some of the

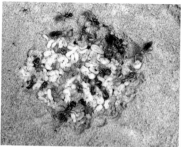

FIG. 3.9. *Left*, close-up of brood in a chamber near the bottom of the nest. Both sexual and worker brood are present. *Center*, a chamber largely filled with sexual brood, and adult males and females. *Right*, the queen attended by a few callow workers. Note the broad midbody, the wing scars, and the large size. Author's photos.

much larger sexual brood. I collected these with a mouth aspirator, for the brood often get injured in the shop vacuum. The density of ants and their average young age indicated that I was approaching the bottom of the nest and the region in which I could expect to find the queen.

Now things got tricky, for I had forgone another deepening of the pit, and the chambers were only a few centimeters above the floor of the pit. There was little room to squat or kneel. I deepened the pit just 10 or 15 cm and took my chances. The large chamber at 240 cm was completely packed with ants and largely sexual brood, and there was hardly any standing room (for the ants) (fig. 3.9, *center*). Ants were tumbling over one another, spreading outward, carrying brood and piling them in every nook and depression. I needed to look sharply for the queen now—she was bigger and had a wide thorax. Aha! There she was (fig. 3.9, *right*). I aspirated her carefully and then aspirated all the workers and brood, clearing the chamber floor of ants.

Was this the last chamber? No such luck. I could see the shaft descending even deeper, so I carried on. Eight centimeters deeper there was another chamber, again packed wall to wall with ants and brood. I aspirated them all, enjoying the sour apple scent, and when the chamber floor was cleared, I saw that there was no shaft to deeper levels. I had reached the end of the nest (fig. 3.10). Whew!

FIG. 3.10. The end of a dig at 2.5 m deep. Frame from a video by Henry M. Tschinkel.

Having excavated the complete nest, I now had to fill in the hole—my least favorite part. I had to move all three metric tons of sand off the tarp and back into the pit, and unlike digging the pit in episodes, filling the pit is pretty much nonstop. I tamped the sand down at intervals so it didn't look too much like a grave when I was done. Looking like a grave is not good. I was once visited by an angry Leon County deputy sheriff who said, "We found what looked like a grave out in the woods, and 'cause there's always missin' persons, we had to dig it up all the way down to six feet. Wadn't nothin' down there but an empty pack of cigarettes. Was that yours?" he asked with a good deal of belligerence. "No sir," I answered. "I don't smoke."

After that, I left my business card on the mounds.

Now I emptied the sand out of my shoes and socks only to realize that I had forgotten to remove my wallet from my back pocket or my cell phone from my left pocket. Since I forget to do this about half the time, my credit cards are sand abraded to the point of refusing to pay for gas at the pump, and my (ultramodern) flip phone crunches and grinds when I open it. As drying sand trickled down my back and sleeves, I took stock of what I had. I had made an annotated acetate tracing of every chamber and collected the contents into a separate bag or box. I made sure to keep all of this out of the sun, as that would

kill the ants quickly. When I counted and measured everything back at the lab, I had a complete census of the colony and its contents, along with the vertical location of all these items—a three-dimensional picture of a harvester ant colony. The sample of chamber tracings in figure 3.2 (*right*) clearly shows how the size and complexity of chambers decrease with depth. My assistant Daniel "Julio" Dominguez turned these tracings into virtual three-dimensional images of nests that slowly twirled and rotated to music, moving the point of view through all points on a sphere (https://www.youtube.com/playlist?list =PLoUFkrkM_ZDcLqVV4TGjDlfplFmWsFJI7).

The meaning of the vertical distribution will become clearer when I discuss division of labor (chapter 8), but for now, my counts produced the vertical structure summarized in figure 3.11. In order to relate my results from this excavation to the results from all my other excavations, I converted all data to the percent total area, dark workers, callow workers, and brood and plotted these against depth. The shaded areas show the approximate range of means for the whole population of colonies at Ant Heaven. You can see that my results for

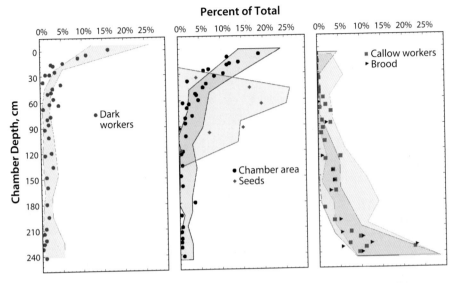

FIG. 3.11. The distribution of dark workers, callow workers, brood, seeds, and chamber area in relation to chamber depth. The points are data from our dig, and the shaded areas show the range of variation (standard deviation) across many nests excavated during the summer.

this excavation are pretty similar—dark workers prevail in the upper nest regions, callow workers and brood in the lower regions together with the brood, and seeds in the upper-middle and chamber area in the upper regions.

OTHER SPECIES

Although we now have a picture of the distinctive architecture and arrangement of a Florida harvester ant nest, it would be good to know whether the nests of other ant species are also distinct. As we will see, most species have characteristic and distinct nest architecture. With respect to the arrangement of the ants within the nest, however, there is less certainty and likely more variation. While most ants seem to prefer to keep their brood in certain regions of the nest, the particular pattern we saw in harvester ants may not be universal. Some ants, like the fire ant, *Solenopsis invicta*, move their brood up and down often, seeking the optimal temperature. In the winter, the sun warms the mound and the ants quickly move their brood there. My student Clint Penick showed that this was in direct response to temperature rather than a daily cycle. He did this by heating the mound at night by placing a grate of glowing charcoal above it. Within a few minutes, the ants were moving their brood into the mound. In the summer, when the soil is warm, the ants don't need to seek higher temperatures in the mound, and the brood is found mostly in the core of the nest.

This kind of heat seeking is quite common among ants, and one can often find brood, especially pupae, piled in windrows beneath warm rocks or logs. Even in harvester ants, pupae are often abundant just beneath the surface, but overall, this is a small proportion of the total brood. It seems possible that the workers "bake" only a specific and brief stage of brood development, for most pupae are found deep in the nest. They may even be manipulating the daily birth rate of adult workers.

Looking through my data from past nest architecture projects, I realized with dismay that although I did a total census of each cast by dissolving or melting it, I did not separate the sections by depth. Therefore, I cannot say whether the trap-jaw ant (*Odontomachus brunneus*), the sandhill carpenter ant (*Camponotus socius*), or any of the three species of long-legged ants (*Aphaenogaster* spp.) host their brood deep in the nest or in some other discrete section. But I did learn from my mistakes, and when I dissolved the five clunky plaster nest casts of the wily fleet ant (*Formica dolosa*) (see fig. 9.6) collecting dust in my garage, I first broke them into five levels. This showed that the *density* of workers, cocoons, and sexuals all increased approximately threefold with depth in the nest. *Formica dolosa*, like *Pogonomyrmex badius*, keeps its brood deep in the nest.

On the other hand, my undergraduate student Tyler Murdock made wax casts of 60 nests of Morris's big-headed ant (*Pheidole morrisi*) and recovered all the ants in 10 cm depth increments. This turned out to be a big job, and I felt a few pangs of guilt about having encouraged Tyler to do this project in addition to his courses. Making the wax casts was a lot faster than analyzing them, and we discovered that they grew mold if stored at room temperature. After that, we stored them in the supercold freezer of the Antarctic Research

Institute along with the ice cores from the Antarctic Drilling Project. We finally put the "ant" in Antarctic. But once completed, Tyler's study was a proof of concept and a milestone in the study of ants and their architecture, for it was the first full study done entirely with wax casting.

Moreover, the study yielded a wealth of information about seasonal cycles, colony composition in relation to growth, nest architectural features, and much more. The location and movement of brood within ant nests in response to conditions and season are important to all ants, and this study revealed these patterns for a particular species. The general pattern was that during the winter *Pheidole morrisi* keep their brood (mostly fully grown worker larvae) in the deeper regions and then move them up between the second and fourth depth decile in the summer, at which time the brood consists of a mixture of brood stages and types. By late summer, the brood have been moved down to the third to fifth decile. In the fall, there are two peaks of brood, one centered in the third decile and another in the eighth and ninth deciles—that is, the bottom of the nest. The soil temperatures at these peaks showed that from spring to early fall, the ants were placing brood at the optimal development temperature of about 23–25°C. Winter placement is more problematical, for most brood are at about 14°C in spite of the higher temperatures deeper in the nest. I suppose the ants know what they are doing—perhaps they are actually regulating the development rate by regulating the experienced temperature. My student Sanford Porter demonstrated that fire ants keep brood at lower temperatures when starved, perhaps reducing the cost of maintenance when growth is reduced. Given a choice of temperatures, the ants can more precisely serve a more subtle goal than maximizing growth rate.

One general feature in common among *Pheidole morrisi*, *Formica dolosa*, and *Pogonomyrmex badius* is that available space is not a controlling factor in the distribution of ants within the nest. In all three species, more than half the chamber area or volume is in the top two or three deciles of the nest, declining to less than 1% to 3% at the bottom of *P. badius* and *Ph. morrisi* nests, and about 8% in *F. dolosa* nests. Yet the density of ants does not track this available space. In *P. badius*, although the number of dark workers was highest in the top regions, their density (workers per cm^2) was five to seven times as high in the deepest regions because chamber area followed the opposite pattern. Callows and brood were more abundant in the lower regions, so their density was far greater there, crammed in as they were. The distribution was similar in *F. dolosa*, with worker and cocoon density at the bottom three times as high as at the top. In contrast, chamber area in the top two-fifths was 60% of the total, decreasing to 8% in the bottom fifth. In *Ph. morrisi*, with its similar distribution of chamber area, ant density was similarly low in the uppermost nest region but peaked in the middle to lower third, depending on the season. In none of these species do the ants seem to be responding to crowding. The relatively huge area of the uppermost chambers in all three species thus presents something of a mystery: Why create these huge chambers when they are so sparsely used? Their function is clearly not to reduce crowding. Indeed, the ants seem to take comfort in crowding, although its role in ant life is largely unknown.

You may wonder about the fate of the harvester ant colony I dug up and censused. Were they done for? You will be happy to hear that they were replanted back where I got them. I gave them a good start on a new nest by constructing an ice nest (see chapter 6) and releasing the colony to occupy this nest. In two or three hours, all the ants were belowground, and a nest disk of excavated soil appeared as workers began enlarging their starter home.

CHAPTER 4

An Inventory of Ant Nests

In the previous chapter, I showed that careful and laborious excavation, level by level, revealed the chamber-by-chamber distribution of the number and types of ants living in a nest. But digging destroyed the architecture. In order to see the full architecture, we need to cast the intact nest into a beautiful rendering of the empty shape, as I described in chapter 2. But casting entombs the ants. The two goals—revealing the architecture and collecting the nest's inhabitants—seem to conflict with one another.

It is possible, however, to satisfy both goals, at least sequentially. It requires choosing the proper casting material. Molten metal is hopeless for achieving both goals, for the hot metal carbonizes the ants so they cannot readily be counted (fig. 4.1). On the other hand, plaster is slightly soluble in water, so after studying, measuring, and photographing the cast, we can break it into smaller

Fig. 4.1. In most casts, the carbonized remains of ants are visible in some chambers. The molten metal presses the ants against the edges and floors. Being wet, the ants are probably not incorporated into the body of the metal, but they are difficult to count in their less than pristine condition. This is part of a cast of the harvester ant *Pogonomyrmex magnacanthus* made in Borrego Springs, California. Author's photo.

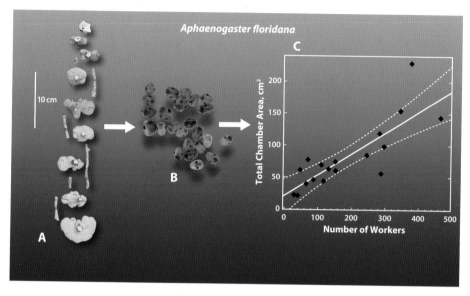

Fig. 4.2. A sample analysis of a plaster cast. *A*, cast of the Florida long-legged ant, *Aphaenogaster floridana*, broken into chambers and shafts for measurement of chamber area, followed by disso-lution of the cast in water. The pieces are in approximate depth order. *B*, the ant heads remaining after dissolution of the cast. *C*, solution of 18 casts of *A. floridana* followed by head counts re-vealed the relationship between number of workers and total chamber area. Dotted lines are 95% confidence bands for the regression. The area increases by 0.3 cm^2 for every additional worker. Author's photos, in part from Tschinkel (2011).

pieces, tie the pieces into fine mesh bags, and put the bags in running water. Over a month or two the plaster dissolves, leaving the ants and other nest con-tents in the mesh bags. Sadly, the ants are usually in pieces, so a census is liter-ally a head count. Still, heads also give information on body size and worker caste, and to a lesser degree, worker age (lighter color). Figure 4.2 shows an example of this procedure for the Florida long-legged ant, *Aphaenogaster flori-dana*, a common ant in the coastal plain pine forests of Florida. The areas of the chambers were measured from the broken cast, which, after being dissolved and sorted, yielded the ants' heads. The graph in figure 4.2 shows the rela-tionship of the number of workers to the total nest area, but many other anal-yses can also be performed, including the distribution of workers and brood within the nest, differences in worker size, and vertical distribution of chamber size and shape. Even though this method is destructive and slow,

the results can teach us much about how ants and their architecture are related, and how the worker population and the nest grow in both relative
and absolute terms. Absolute nest size is measured in volume (milliliters or
liters), and absolute population size is measured by count, but the two do not
necessarily grow at the same rate. Dividing the nest volume by the number
of ants produces the relative size, for example milliliters of nest per ant. If
the population grows at the same rate as the nest volume, then the amount
of space per ant remains constant no matter how large the nest. If the population grows faster, then the ants in large nests are more crowded than those
in small nests, and of course the reverse can also occur. The role of crowding in ant nests is unknown, but it is easy to imagine that it is important, as
it probably affects social life.

A less cumbersome method for assessing the worker population and its distribution is to make the cast with molten paraffin instead of plaster. Wax casts
are too fragile for serious study of the architecture or for display, but in contrast to dissolving plaster casts, wax casts can be melted to recover all nest contents intact, even fragile larvae and pupae, and any other creatures that shared

FIG. 4.3. *Left*, a wax cast of a nest of Ashmead's long-legged ant, *Aphaenogaster ashmeadi*. *Right*,
the ants recovered upon melting the cast. All nest contents are recovered completely intact.
Author's photos, in part from Tschinkel (2010).

the nest or objects the ants collected (fig. 4.3). Combining metal casting of some nests in a set with wax casting of others is a good way to satisfy both census and architecture goals.

CREATING A SPECIES INVENTORY

Having gained proficiency and skill in melting zinc and aluminum, I set out to make nest casts of as many ant species as possible, mostly the larger species nesting in the sandy soils of the piney woods south of Tallahassee. Table 4.1 lists the scientific names, abbreviations, and common names of all the species I have cast. Finding the nests of many of these larger ant species requires little skill—workers dump the excavated soil into easily visible heaps of pellets on the ground surface. The shape and size of the dumps and the size of the pellets even identify the ant species (fig. 4.4)—Florida harvester ants (*Pogonomyrmex badius*) make large flat disks that they cover in bits of charcoal, perhaps for aesthetic reasons; the fungus-gardening ant *Trachymyrmex septentrionalis* creates half crescents like a Turkish flag to one side of the nest entrance; some species of cone ants, *Dorymyrmex*, make perfectly symmetrical craters like tiny volcanoes; and workers of the sandhill carpenter ant, *Camponotus socius*, scatter coarse pellets over a square meter, like a strewing pebbles upon the water. A few make distinctive mounds—Morris's big-headed ant, *Pheidole morrisi*, the commonest ant in the piney flatwoods, makes neat, conical mounds like squat traffic cones of sand, tiny bits of plant matter, and charcoal, with a few chambers contained in the mound. The exotic fire ant, *Solenopsis invicta*, makes large mounds honeycombed with sinuous galleries in which the brood are warmed on cool, sunny days. Workers of the Florida long-legged ant, *Aphaenogaster floridana*, are a bit shyer—they initially make distinct soil dumps around their nest openings that become indistinct after rains, but the arrangement of little sticks around the openings to form squares, like log cabins, still identifies the owners of the nest.

But there are many much smaller ant species whose nests range from inconspicuous to devilishly hard to find (fig. 4.4). Locating them relies on the

TABLE 4.1. Scientific and common names of the ants on which this book is based

SCIENTIFIC NAME	ABBREVIATED SCIENTIFIC NAME	COMMON NAME
Aphaenogaster ashmeadi	Aph. ashmeadi	Ashmead's long-legged ant
Aphaenogaster floridana	Aph. floridana	Florida long-legged ant
Aphaenogaster treatae	Aph. treatae	Treat's long-legged ant
Camponotus socius	C. socius	Sandhill carpenter ant
Cyphomyrmex rimosus	Cy. rimosus	Immigrant little fungus gardener
Dolichoderus mariae	Dol. mariae	Mary's tongue-and-groove ant
Dorymyrmex bureni	Dor. bureni	Buren's cone ant
Formica archboldi	F. archboldi	Archbold's fleet ant
Formica dolosa	F. dolosa	Wily fleet ant
Formica pallidefulva	F. pallidefulva	Variable fleet ant
Monomorium viridum	Mo. viridum	Metallic trailing ant
Myrmecocystus kennedyi	My. kennedyi	Kennedy's honeypot ant
Myrmecocystus lugubris	My. lugubris	Gloomy honeypot ant*
Myrmecocystus navajo	My. navajo	Navajo honeypot ant*
Nylanderia arenivaga	N. arenivaga	Sand-loving crazy ant
Nylanderia parvula	N. parvula	Northern crazy ant
Nylanderia phantasma	N. phantasma	Ghostly crazy ant
Odontomachus brunneus	O. brunneus	Southeastern trap-jaw ant
Pheidole adrianoi	Ph. adrianoi	Rosemary big-headed ant
Pheidole barbata	Ph. barbata	Bearded big-headed ant*
Pheidole dentata	Ph. dentata	Versatile big-headed ant
Pheidole dentigula	Ph. dentigula	Woodland big-headed ant
Pheidole morrisi	Ph. morrisi	Morris's big-headed ant
Pheidole obscurithorax	Ph. obscurithorax	Large imported big-headed ant
Pheidole psammophila	Ph. psammophila	Sand-loving big-headed ant*
Pheidole rugulosa	Ph. rugulosa	Rough big-headed ant*
Pheidole xerophila	Ph. xerophila	Dry-loving big-headed ant*
Pogonomyrmex badius	P. badius	Florida harvester ant
Pogonomyrmex californicus	P. californicus	California harvester ant
Pogonomyrmex magnacanthus	P. magnacanthus	Little harvester ant*
Prenolepis imparis	Pr. imparis	Winter ant
Solenopsis geminata	S. geminata	Tropical fire ant
Solenopsis invicta	S. invicta	Imported fire ant
Solenopsis nickersoni	S. nickersoni	Nickerson's thief ant
Solenopsis pergandei	S. pergandei	Pergande's thief ant
Trachymyrmex septentrionalis	T. septentrionalis	Tuberculate fungus gardener
Veromessor pergandei	V. pergandei	Desert black harvester ant

Note: Many of the common names are taken from Mark Deyrup, Ants of Florida (Boca Raton, FL: CRC Press, 2017).

*In a few cases that lacked a common name, I made one up, as allowed by international treaty.

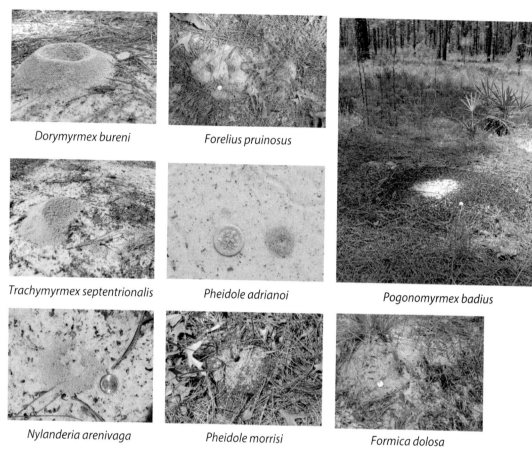

Dorymyrmex bureni

Forelius pruinosus

Pogonomyrmex badius

Trachymyrmex septentrionalis

Pheidole adrianoi

Nylanderia arenivaga

Pheidole morrisi

Formica dolosa

FIG. 4.4. A sample of the nest disks of ant species found at Ant Heaven. In each image, a US dime provides scale. Author's photos.

fact that ants are always hungry, and most species recruit nestmates to help retrieve food. A few cookie crumbs on a card will soon draw a trail of tiny recruits that can be followed, often with difficulty, back to the nest.

If there is a lot of leaf litter, following the trail can be tedious to the point of giving up, so it helps to rake the litter off a few square meters a couple of days ahead of baiting. The first time you notice that what looks like an empty bait card is actually occupied by the tiny blond workers of the woodland big-headed ant, *Pheidole dentigula*, or Nickerson's thief ant, *Solenopsis nickersoni* (fig. 4.5), you will be astounded by how small they are, hardly larger than a single number in the date on a dime. Theirs is a tiny universe hidden in a square-meter patch of leaf litter. Their trails can be the equivalent of you walking all of Manhattan north to south, carrying your brother on your back. Sometimes their nests are marked by soil dumps of tiny pellets, not even covering a dime, and sometimes the workers simply disappear into an opening so small it looks like simply a space between sand grains. This is the moment when you reflect on the scales at which we and other creatures live.

Even among ant species, life plays out at a wide range of scales. A nest of rosemary big-headed ant, *Pheidole adrianoi*, whose black workers look like walking specks of dust, may be a few centimeters from a nest of the sandhill carpenter ant, *Camponotus socius*, whose workers weigh over a thousand times as much, and instead of ranging a meter from the nest to forage, they range dozens of meters. The world of the former would fit on a small tabletop, while that of the latter would occupy the footprint of a typical suburban house. Do their worlds actually intersect, or are they merely layered in the same physical space? The general importance of body size in natural history suggests that the tiny ants and the large ants occupy different sections of the ecosystem, exploit-

FIG. 4.5. Nickerson's thief ant, *Solenopsis nickersoni*, with a US dime for size comparison. These are among the smallest ants in the southeastern United States. Author's photo.

ing different resources and living together peaceably, perhaps aware of each other's presence simply as an annoyance, like people and flies. It is unlikely that tiny ants live peaceably with other tiny ants, for to varying degrees, they must exploit similar resources and thus come into competition. But although ecologists are fond of the notion of competition, this is just theory, for no one has done the necessary experiments to test these ideas.

I followed the trails of these tiny ants from baits back to their nests so I could try to make casts of them. These nests should scale to the body sizes of the ants—that is, their shafts and chambers should be tiny. Because heat is lost only across surfaces, the rate of loss is proportional to surface area. On the other hand, the amount of heat that can be lost is proportional to volume, so any molten metal poured into a tiny ant nest with long, skinny shafts will cool and freeze extremely quickly because the surface area is so large compared to the volume. Therefore, the lower the metal's melting point and the hotter the melt, the farther it will flow before freezing. Aluminum is a complete dud for tiny nests and usually does not penetrate them at all. Zinc can be made to work, but for such tiny nests, even zinc has to be herded, sluiced, and encouraged to flow down these narrow shafts. I use an aspirator to suck away sand to expose the top of the shaft and shape it into a narrow, conical funnel with the shaft at the lower vertex, readying the nest to receive the molten zinc.

I begin by melting a small amount of zinc in a graphite crucible in a small kiln. As with aluminum, the hotter the melt, the deeper it will flow down the nest, but there is an upper temperature limit at which zinc ignites and burns, forming a wispy zinc oxide smoke that makes you cough and forms deep cushiony deposits in the crucible that look like cotton candy (fig. 4.6, *left*). Not surprisingly, this interferes with pouring the molten metal, but the beautiful chartreuse fire makes it almost irresistible to ignite the zinc at least once during every day out. A bit of cooling and a spoon to scoop out the oxide put the zinc back in working order, no harm done. However, letting the kiln reach extreme temperatures can result in disaster with metal crucibles because the heat burns the crucible from the outside and the metal dissolves it from the inside (fig. 4.6, *right*).

FIG. 4.6. *Left*, a crucible of zinc that was overheated and caught fire, emitting dense white zinc oxide smoke. *Right*, a crucible that was destroyed by the solvent power of molten metal and the oxidative action under extreme overheating. Author's photos, from Tschinkel (2010).

When it is time to pour, accuracy is very important, for the first splash of molten zinc must be directly into the conical funnel made earlier. The zinc must then be poured just fast enough to keep this cone filled so that the pressure of the molten zinc keeps the metal flowing downward. You can see the little zinc puddle deflate as the metal drains downward. In the tiny nests of *Pheidole adrianoi* or the sand-loving crazy ant, *Nylanderia arenivaga*, you are doing well if the metal flows down at least 25 cm before freezing. The entire cast is frozen in no more than two or three minutes, when it is time to dig a pit next to the cast, expose it from the side, and remove it, preferably without breaking it.

Finding where the cast ends and the nest continues is not always easy—an aspirator helps remove sand from the continuation of the shaft to the rest of the nest. We repeat the formation of a conical "entry funnel," pour, and dig again. And again, the second piece may not complete the cast. Lord help you if you need a third or fourth casting, because maneuvering a glowing crucible in a narrow pit is tricky.

How do we know when the cast is complete? This is actually easier than it would appear—the lower end of the metal freezing in midshaft freezes in air, not in contact with sand, and it has a smooth, rounded contour, whereas metal freezing in the actual end of a nest, be it shaft or chamber, bears the pebbled surface of the sand in which it froze. The only time this fails is when a shaft is blocked by fallen sand, but even then, the metal tends to end in recognizable disorganized steps. Should this be the case, it is time to launch a search for the continuation of the shaft.

QUEEN OF NEST ARCHITECTS: THE FLORIDA HARVESTER ANT

After casting the nests of a couple of dozen ant species, I had definite favorites, chief among them the Florida harvester ant. Even if you don't know anything about its subterranean nest architecture, the Florida harvester ant is a charismatic ant. Its large, charcoal-covered nest disks contrast with the white soil of the sandhill longleaf pine–wiregrass forest. The uninitiated usually refer to it as "the red ant" about "yea big," indicated between thumb and forefinger, unaware that about 80% of the area's 100 ant species are "red ants," and that by ant standards, harvester ants are large. For me, the harvester ant offers many attractions—it is conspicuous and large enough so that much of its behavior is observable without a microscope, and it collects seeds, an interesting and unusual behavior. What is more, like me, it is active only in the daytime and is deliberate in its habits, and it generally offers many interesting questions begging for an answer—in other words, a myrmecologist's dream ant. Knowing about its beautiful and elaborate nest architecture, as described in the previous chapters, adds enormously to its charismatic nature.

Like me, most people find ant nest casts aesthetically pleasing—the linear simplicity of the Florida long-legged ant (*Aphaenogaster floridana*) or Buren's cone ant (*Dorymyrmex bureni*); the tight, dense packing of the metallic trailing ant

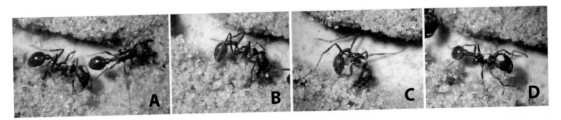

FIG. 4.7. Sequence of actions by Florida harvester ants during the formation of soil pellets. *A*, shearing and cutting with the mandibles; *B*, pushing sand backward; *C*, curling the body to compress sand into a pellet; *D*, carrying the pellet between the mandibles and the psammophore. Author's photos.

(*Monomorium viridum*); the complex confusion of the fire ant (*Solenopsis invicta*); the sheer enormity of a leafcutter ant (*Atta* sp.) nest (see fig. 9.2)—but for all but a few people, the Florida harvester ant (*Pogonomyrmex badius*) is the Queen of Nest Architects. The first of its many features to command one's attention is the sheer size of the nest, actually human in scale (fig. 2.2). A fully mature colony of 5,000 to 10,000 ants builds a nest as deep as a human is tall, or deeper (1.8–2.0 m). The deepest I have ever excavated resulted in a pit so deep that a person standing on the shoulders of another would barely see over the edge (3.1 m). Rarely can I see over the edge of the pit that results from digging out a typical cast (fig. 3.10).

Sheer volume goes along with depth. Casting the nest of a mature colony usually requires my tallest crucible full to the rim with molten aluminum. The largest nest I ever excavated measured about 11 liters in volume. This extravagance of space begins in the colony's infancy—a colony with 8 tiny workers builds a nest 30 cm deep and 50 ml in volume, about 2,000 times the combined body volumes of the ants. At 1,000 workers, it has grown to about a liter, or 150 times their combined body volumes, and by 5,000 workers, it has attained a mean of 5 liters, or 125 times their volumes, and a depth of 2 to 2.5 m.

Workers have loosened all this volume of sand from chamber edges and shaft ends, formed it into pellets one at a time, and carried it upward in stages to be dumped on the nest disk above. The workers bite into the sand wall to loosen

Top view

Bottom view

FIG. 4.8. Chambers of the Florida harvester ant from just below the surface down to about 10 cm are very complex and are probably formed by the widening and coalescing of branching, horizontal tunnels. The top image is from above, the bottom from below. The nest's two entrances, where molten metal was exposed at the surface, can be seen in the top image. Note the shaft with the first simpler chambers at the upper right of each image. Author's photos.

the grains, push them backward with the mandibles until they are under the body, and then flex the abdomen forward to compact the grains into a pellet, which they then pick up and carry (fig. 4.7). Pellets can be formed only in damp sand, as the thin film of water causes the grains to stick together.

Sand has a bulk density of 1.5 kg per liter, and a large nest has a volume of, say, 10 liters, so the ants excavate 15 kg of sand to form the nest. A sand pellet at Ant Heaven contains an average of 160 sand grains and weighs about 9 mg (more than the weight of the ant carrying it), so we can calculate that the 15 kg

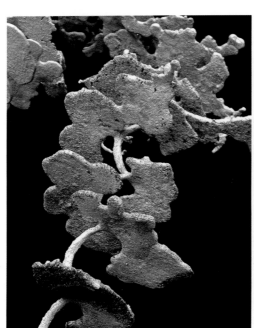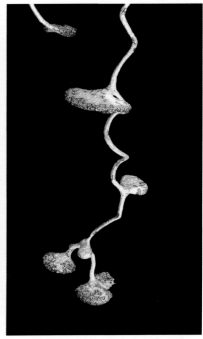

FIG. 4.9. *Left*, deeper chambers become farther apart, less complex, lobed, and smaller. *Right*, they become simple kidney shapes in the deepest regions. The helical nature of the shaft is especially apparent in the right image. The helices can wind either clockwise or counterclockwise, even within the same nest. Author's photos.

of sand was transported in at least 1.67 million pellets, each requiring at least one trip to the surface. For every liter of nest volume, the ants carried pellets upward for a total of 60 km, so our large nest was the product of 500 to 600 *kilometers* of upward travel, the distance from Philadelphia to Cleveland. And remarkably, all of this was accomplished during a nest relocation in four to six days, as I will reveal in the next chapter. Nothing in the world of these ants is modest or moderate. To think otherwise is merely a failure to see the world at their scale.

But as impressive as the nest size and work effort are, it is the beauty of the architecture that is most appealing (fig. 2.2). Just below the top surface lies a huge lacy meshwork of interconnected tunnels (fig. 4.8), the largest

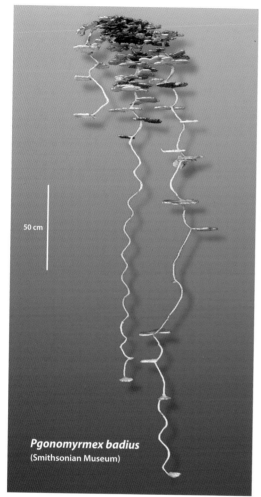

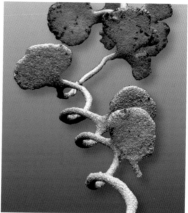

Pgonomyrmex badius
(Smithsonian Museum)

50 cm

FIG. 4.10. *Left*, a harvester ant nest cast showing a long, helical shaft without chambers. Note the different lengths of the shafts. *Right*, chambers are almost always formed on the outside of the helix. Author's photos, in part from Tschinkel (2015a).

horizontal extent of the nest. This complexity becomes simpler and smaller in area with depth. By a depth of 15 to 20 cm, there is no longer a meshwork of chambers but rather single chambers with complex, deeply lobed perimeters (fig. 4.9). At this depth, the helical nature of the shaft connecting the horizontal chambers is apparent and becomes one of the most conspicuous features of the remainder of the nest (fig. 4.10). There seems little preference for the direction of the helix, for even within a single nest it may sometimes

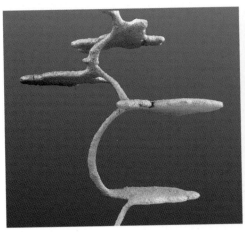

5 cm

FIG. 4.11. *Left*, chamber floors are always smooth and perfectly horizontal. *Right*, shafts in the upper nest region are usually wide and ribbonlike to accommodate the higher ant traffic, much like multiple lanes on a superhighway. Photos by Charles F. Badland, in part from Tschinkel (2004).

wind clockwise and then switch to counterclockwise. Deeper and deeper, the chambers become simpler and simpler, smaller and smaller, and farther and farther apart. Below about 40 to 50 cm, chambers have a simple kidney shape and always lie connected to the outside of the helical shaft, like flat grapes on a helical vine (fig. 4.10). Perhaps accommodating the size of the ants, chambers average about 1 cm from floor to ceiling, their floors always, always perfectly horizontal, smooth as a ballroom floor (fig. 4.11). Clearly, the ants have precise knowledge of up, down, and horizontal. They can probably sense smoothness, too, although what looks smooth to us may well be rough to the ants. At the ants' scale there is clearly a limit to how smooth a granular material like sand can be made. Shafts, too, average about 1 cm in diameter. Near the surface they are larger to accommodate the increased traffic and are often wide, flattened, and ribbonlike (fig. 4.11), the multilane interstate highways of the nest.

Is this beauty important to the ants? How would we possibly know? Science cannot ask such questions. Science postulates that this beauty has a practical purpose, and indeed, among the many definitions of beauty is this one from the *Oxford English Dictionary*: "That quality or combination of qualities . . . which charms the intellectual or moral faculties, through inherent grace, *or fitness to a desired end* [my emphasis]." That is, it is suited for a purpose or function. Why do the ants make a helical shaft when a vertical one would get them there faster and be less work? Is it because a seed dropped in a vertical shaft would fall all the way to the bottom? Perhaps, but we will see below that colonies offered a perfectly vertical, straight shaft usually accept it without complaint, festooning it with chambers at normal intervals. So do the ants like helical shafts for their beauty? We hardly know how the architecture fulfills particular practical purposes, so the question simply hangs in the air. I am satisfied to let it hang there, though that will not stop me from trying to figure out how the architecture serves the colony that built it. I am aware that science is gradually chipping away at the mystery of beauty, but except for a few frayed edges, the mystery is still largely intact, as the nests of *Pogonomyrmex badius* attest.

THE SCALE OF ANTS AND THEIR NESTS

One of the gifts that a study of tiny creatures can offer is the awareness of scales of existence, and how different life is at different scales. The data I have collected on the sizes of nests and the sizes of the populations that built them invite us to think about the nature of these accomplishments in relation to their scale (size). There are several scales to consider—the total worker population and its nest, the individual worker and her achievements, and the individual worker in relation to the material with which she is working.

It is striking how impressively deep the nest of the rosemary big-headed ant (*Pheidole adrianoi*) is, considering the tiny size of the ants (fig. 4.12). Our judgment of impressiveness is probably biased by the much greater size difference between ourselves and the tiny *Ph. adrianoi* than between *ourselves* and the much larger *Pogonomyrmex badius*. Is the 10 liter, 250 to 300 cm deep nest of *P. badius*

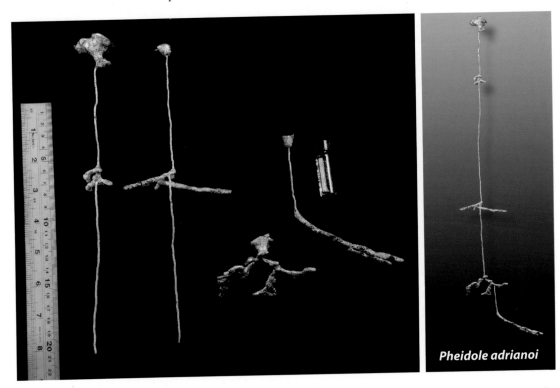

Fig. 4.12. *Left*, this zinc cast of a nest of the rosemary big-headed ant, *Pheidole adrianoi*, required four sequential pours, left to right. *Right*, the pieces assembled into a complete cast. The four complex chambers are connected by a fine vertical shaft. Nests at this fine scale are hard to cast. Author's photos.

more or less impressive in relation to the size of the ants than the 60 cm deep nest of *Ph. adrianoi*? Such relationships are described as scaling, or allometry—how fast a measure increases as the size of an animal increases. Just as an example, our heads grow more slowly than our bodies, which is why babies look like babies with their big heads and short legs, and why grown-ups look like grown-ups with relatively smaller heads and longer limbs. As another example, metabolic rate increases more slowly than body size in mammals. If we had the same metabolic rate per gram as a shrew, we would probably ignite in still air. Strength also scales much less than one to one, so we are astounded at the relative strength of an ant. In truth, if we were the size of ants, we would be just as strong. If elephants could talk, they would be impressed by *our* relative strength

FIG. 4.13. *Left*, the size difference between the rosemary big-headed ant, *Pheidole adrianoi*, and the Florida harvester ant, *Pogonomyrmex badius*, is very large—1.4 mm vs. 6 mm in length, and 0.04 mg vs. 4 mg in dry weight, a factor of 100. Both are shown at the same scale (photos by April Nobile, AntWeb.org, modified by Walter R. Tschinkel). *Right*, both species on a US dime. The *Ph. adrianoi* worker is just to the right of "God" and above "trust." Author's photo.

and incapable of comprehending that of an ant. No elephant can lift its own weight, and although many humans can, ant workers can lift many times theirs.

Applying the principles of scaling to nest architecture raises a question: Does the relative nest-excavating prowess of ants change with ant size? Many biological measures increase at different rates than the size of the animal, so perhaps the relative size of the nests can be expected from the relative size of the ants (fig. 4.13). Over the years, I have accumulated the necessary data for making such comparisons (nest volume, worker body weight, and worker number) for ten species of ants.[1] All of these measures vary enormously among these species—for example, *P. badius* workers are about 100 times as heavy as those of *Ph. adrianoi* (4 mg vs. 0.04 mg, dry weight), their nests have 600 times the volume (3,000 ml vs. 5 ml), and their colonies contain 15 times as many workers (3,000 vs. 200), resulting in a total colony weight that is 1,500 times as large (12 g vs. 0.008 g).

---

1 *Pogonomyrmex badius, Pheidole adrianoi, Ph. morrisi, Camponotus socius, Aphaenogaster floridana, A. treatae, A. ashmeadi, Odontomachus brunneus, Formica dolosa, F. pallidefulva*

Not surprisingly, ant species with larger workers (2 to 7 mg) excavate a larger volume per worker (1 to 4 ml). Those with medium-sized workers (0.5 to 2 mg) provide about 0.3 to 0.5 ml per worker, and those with tiny workers (0.04 mg to 0.2 mg), about 0.02 to 0.06 ml per worker. Simply put, larger workers need more space and are capable of excavating it. Thus, volume per worker does not provide a fair comparison of nest-digging prowess. To compensate for differences in worker size, it is better to use the volume excavated per gram of workers, as this will eliminate the effect of worker size and put all colonies on the same size footing. Measured this way, P. badius excavates about 250 ml for every gram of workers, whereas Ph. adrianoi excavates 600 ml per gram, making its tiny nests a relatively greater feat than the giant nests of P. badius.

Does this mean that the nest-digging prowess per unit weight of workers declines with worker size, making the feats of tiny ants more amazing than those of large ants? Unfortunately, data for my 10 species are insufficient to give a clear answer. The volume per gram of workers ranges from about 120 to 1,300 ml per gram, but it follows no clear patterns with respect to worker or colony size. Seven of the ten species dig between 100 and 300 ml per gram, and three species dig more than 500 ml per gram. Of course, all of these represent a very high degree of nest-building prowess. If I were to dig a nest of 300 ml per gram of my dry body weight, I would have to excavate 13,000 liters, or about 20 metric tons of sand. Just thinking of it makes me tired.

Finally, there is another important scale of interest, that of the individual ant in relation to the granularity of the soil it excavates. As the size of an individual worker ant decreases, its size approaches that of a soil grain. In the sandy soil of my favorite sandhill research site (Ant Heaven), a single sand grain is the smallest unit that can be carried. Because it is quartz, it weighs about 2.4 times as much as an ant of the same volume. Thus, the average sand grain at Ant Heaven is about half the volume but 1.4 times the dry weight of a worker. So an average man would have to carry rocks that averaged about 115 kg. To put things into the proper perspective, Ph. adrianoi constructs its nest in a pile of boulders (fig. 4.14, left).

FIG. 4.14. *Top left*, the head of a *Pheidole adrianoi* worker shown in relation to the sand in which it nests. *Top right*, minor and major workers in the sand in which they excavate their nests. The average sand grain weighs 1.4 times as much as a minor worker—the ants are essentially nesting in a pile of boulders. *Bottom*, *Pogonomyrmex badius* nesting in colored sand. The average sand pellet contains 160 grains. Workers weigh 70 times as much as the average sand grain. The coarseness of the medium is much greater for *Ph. adrianoi* than for *P. badius*. Author's photos.

With an increase in worker size, the relative granularity or coarseness of the soil decreases, so *Ph. morrisi* workers (0.18 mg) are about three times as heavy as the average grain, and *P. badius* workers (4 mg) are 70 times as heavy. For efficiency of transport, most ants form damp soil into obvious pellets of multiple grains. For example, an average *P. badius* pellet contains 160 grains and weighs 9 mg, more than twice the weight of the worker carrying it (fig. 4.14, *bottom*). Whether pellet size is optimized by ants is an open question.

Humans are understandably "scale chauvinists"—we live in a world of limited, but much larger scales than the world of ants. I have tried to relate the feats of ants to our scale, but of course this ultimately misses the point because the feats of ants and the feats of humans are all scale dependent. If we were the size of ants, their feats would seem ordinary to us and we would be able to match them without a problem.

We now have a number of examples of the finished products that ants create, as well as the scale of their achievements. But through what processes do the ants excavate their nests? Does the colony simply enlarge the founding nest, or does it dig new nests and move into them? On what timescales does this take place? In the next chapter, I will describe these processes and rates for the Florida harvester ant.

CHAPTER 5

Moving House

Ants expend so much time and effort in excavating a nest, it would seem reasonable for them to live in it for a long time, maybe even for the entire life of the colony. Indeed, for much of the history of ant studies, colonies have been regarded as more or less permanently rooted in one place, much like plants. Ecologists have even pointed out the plantlike characteristics of ant colonies— rooted in place, they interact only with their neighbors, drawing all their resources from their fixed neighborhoods; like plants, they grow by adding modules (workers or leaves/stems) and shrink by shedding modules; finally, they reproduce by emitting "propagules" (seeds or mated alate queens). As interesting as these comparisons are, by the early 1980s, reviews of the scientific literature showed that ant colonies were much less fixed in space than they had previously appeared.

For ants that nest in preformed cavities such as hollow twigs, acorns, nuts, or rotting logs, the deterioration of the nest seems a plausible reason for moving to better quarters. For these ants, frequent moves come as no surprise and involve little time or risk. After all, the forest floor abounds in hollow twigs, empty seedpods, and other available cavities, so ant colonies need not put up with nest deterioration. Such nest relocations are easy to study in the laboratory—one simply lifts the cover off the nest and watches the panicked ants choose one of the offered alternative nests, noting the choices they make, how quickly they move, and how they organize the move. We therefore know a good deal about such moves and how the ants make the necessary decisions.

But for ants that have expended the enormous labor of digging a nest in soil, such reasons are harder to conjure up. And yet move they do, once again investing a huge amount of time and effort in excavating a new nest and accepting the risk of predation, desiccation, and overheating that comes with exposure during an aboveground move. Even the huge colonies of the leafcutter ants (*Atta* and *Acromyrmex* spp.) have been known to abandon their massive nests, though very rarely. For most of the studied species of ants, there is no obvious (or even less than obvious) reason for such moves, but since they do move, we have to assume that moving is favored by natural selection, thus maintaining this behavior in the population.

HARVESTER ANT COLONIES MOVE OFTEN

When I started digging up Florida harvester ant colonies in the mid-1980s to study their development and seasonal natural history, I already knew that their colonies moved often—a visit to a previously flagged nest often revealed only a sickly ghost of a charcoal disk, thinned and dispersed by wind and water (fig. 5.1). No ants bustled about on its surface, dumping sand pellets or trash. Did it die, was it inactive, or did the ants simply move? A brief scouting often revealed a fresh disk nearby, replete with charcoal and home to a vigorous population of ants going about their quotidian tasks. Since all the other neighboring colonies were at their previously flagged locations, there could be no doubt that the unmarked active and marked inactive nests both belonged to the same colony, and that the colony had moved.

Whatever the reason for moving, moves offer an opportunity to study the creation of a nest from scratch. The entire beautiful, complex architecture of *Pogonomyrmex badius*, as well as the arrangement of contents within the nest, is

FIG. 5.1. Florida harvester ants move often, but usually not far. The charcoal disks of their vacated nests are visible as "ghosts" for many months or even years. The current live nest is on the left. Author's photo, from Tschinkel (2014).

created in only a few days as the colony moves from its old nest to a new one. The move consumes a mere moment in the colony's total lifetime. The colony's footloose nature makes it an excellent subject for studying the nest's architecture. We come to understand that the nest is not the product of the gradual enlargement of the founding nest but is created over and over during the life of the colony. A study of nest relocation is thus a starting point for understanding nest architecture as well as other aspects of the ants' biology.

Colonies give no obvious warning that they are going to move, making it impossible to predict when a *particular* colony is going to move. Therefore, to get a large enough sample size to study relocation, we need to track a whole population of colonies so we can catch some in the act of moving. I had already invested a good deal of time and effort in the *P. badius* population at Ant Heaven, so the choice of population was natural. As with many projects in science, I am telling the story backward—it is not that I had the idea of studying the principles of nest relocation and chose *P. badius* as the most suitable for this project. It was actually the frequent moves of *P. badius* at Ant Heaven that gave me the project idea in the first place.

Tracking the population

The first step in this population study was to give each colony in this 23-hectare area a number, an identity. A search of the "heart" of Ant Heaven in 2010 yielded over 200 colonies, each marked with an aluminum tag on a wire stake so our tags would survive the occasional prescribed burns by the Apalachicola National Forest. In 2011, we expanded the surveyed area to the full extent of Ant Heaven, creating an inventory of between 400 and 450 numbered colonies. Many of the originally marked colonies are still present as I write this over a decade later.

Although a map was not essential to studying relocation, mapping added the highly desirable ability to analyze the data spatially—that is, determine not only how far and in which direction each colony moved, but also its relationship to its close and distant neighbors, and to its environment. At this point, GPS technology lent a hand. I bought a used Trimble GeoXT GPS receiver on eBay for less than a quarter of the (breathtaking) new price. As it was supposed to be capable of specifying a location on Earth to within 50 cm, it was certainly up to the job. Setting the GPS to record positions every second and accumulating a couple of dozen such readings resulted in a reasonable compromise between accuracy and time. The calculated average location could be off from the true position by a few meters, errors that varied over time, but even this could be corrected by reference to a nearby "base station" whose latitude and longitude was known to within a centimeter or two. Base station data are available online, so for example, at 3:25 p.m. on May 14, 2011, the satellites were reporting base station locations 1.6 m northeast of its true location (in latitude and longitude). This error also applied to my colony locations, so correction required simply subtracting the latitude/longitude error from the device-reported location. The upshot was that the location of the vast majority of my colonies was identified to within 50 cm on the face of a planet that is 12,756 km in diameter and is not even a perfect sphere, as it is flattened 43 km by the centrifugal force of its rotation.

With the whole population numbered and precisely located, it was simple, at least in principle, to keep track of nest relocations—we had to visit each colony at intervals to determine whether it had moved, and where. If it had moved, we recorded its disk diameter and new location on the GPS and moved its numbered tag and flag to the new location. We settled into a routine of checking each colony five to seven times a year, adequate for the observed nest relocation frequency. At each visit, we recorded whether the colony was active or inactive, had moved or not moved, or was a new, as yet unrecorded colony. This produced detailed information about moves (which we also used to estimate colony life spans; see chapter 8). Over six years, my field assistants Julio Dominguez, Nicholas Hanley, Tyler Murdock, and Neal George always looked forward to wandering the pleasant woods of Ant Heaven and visiting colonies whose numbers, locations, and antics they often remembered.

Another satellite-based technology allowed us to present our maps of colony locations on actual images of Ant Heaven, as if seen from heaven. The precise GPS latitude and longitude were added as place marks in Google Earth, yielding a bird's-eye view of our population (fig. 5.2). Each survey was added as a "layer" in Google Earth, and the colony symbols were coded for size, date

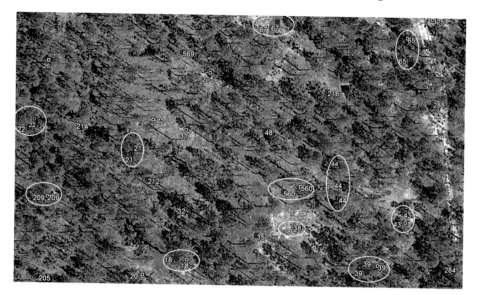

FIG. 5.2. A small section of Ant Heaven on Google Earth, showing the locations of colonies during 2013. Each different-colored symbol shows the location in a different survey. Ellipses enclose the locations of colonies that moved during 2013. From Tschinkel (2014).

of survey, and anything else I had the patience for. Layers could be turned on or off. Printouts of each survey made searching during the next survey easier, because each colony was crossed out once it had been found, and no colonies were missed.

So, with the help of this multibillion-dollar system first designed for precise targeting of nuclear missiles from submarines, we could now pinpoint and map the location of every one of our colonies with the precision of an incoming ICBM (actually better).

When, where, and how often do they move?

So what did we learn? First and most obviously, colonies move a lot. They are, taken all together, a restless lot, but some are definitely more restless than others. At the complacent end were colonies that moved only once in three years, while on the neurotic end were some that moved four times in a single summer. These differences in "personality" (a trendy term for consistent differences) were not associated with any trait I could detect. The average number of moves for all colonies hovered around once a year (0.72 in 2012 and 1.15 in 2013). You might think that because of the work involved in digging a new nest and moving into it, colonies would move a substantial distance to make it worthwhile, but you would be wrong. At the lazy end were colonies that moved only a meter or so, digging the new nest so close that even an ant standing on the old disk rim could see the new one. At the more energetic end were a few colonies that established their new nest more than 10 m away (up to 40 m, the equivalent of a human moving 8 km, the length of Central Park and back). But the average for all the colonies was a mere 4 m, more or less like moving down the street.

Is there a preferred direction of moving? In the population as a whole, colonies moved in random directions. But that does not mean that individual colonies did not have a preferred direction in which they repeatedly moved. Nevertheless, each move in a sequence of moves by individual colonies zigged and zagged randomly, so the more a colony moved, the closer to its original location it ended up—colonies were doing a random walk around a point, like Brownian motion. Moving only once in two years placed a colony about 4 m from its starting point (of course), whereas moving six times in two years and traveling a total of 30 m placed the colony less than 1 m from where it started. This suggests that the ants were moving not to get away from or closer to anything, but for some other reason that is still a mystery.

Moving was highly seasonal. Very few colonies moved between about November and late May, but in mid-June moves began in earnest, peaking in July and August, when about 1% of colonies moved every day, and 25% to 35% moved during each of these months. To stroll through Ant Heaven during this time of year invariably revealed one or more colonies in midmove, with streams of workers running in both directions on trails connecting old and new nests.

Why do they move?

So why do they move so much? Are they seeking a better neighborhood? Have they outgrown their current house and need a bigger one? Are their ant guests becoming a tiresome bother and they are trying to leave them behind? Is the nest getting contaminated with nasty diseases, or just filling up with trash?

It seems unlikely that colonies improve their neighborhood by moving—they move only 4 m, but their closest neighbor is 16 m away, and the next one 18 m. They are in fact boxed in by neighbors all around, and to move through a neighboring territory would be akin to Napoleon's march through Russia. Not a good idea.

Perhaps, for reasons known only to the ants, some parts of Ant Heaven are more conducive to a settled life than others. As a result of different management histories including logging, planting, and burning, some parts do indeed have different canopy coverage, tree species composition, and litter density. Nevertheless, there was no difference in the rates of moving among any of these blocks.

Maybe the ants need a bigger nest because the colony has grown? To check this hypothesis, we dug up, censused, and mapped both the old colony (now vacant) and the new colony a week or two after the move. The new nest was an awful lot like the old nest (fig. 5.3). It was just a little smaller and not quite as deep, but of course the ants were still working on it. The vertical distribution of chambers, ants, brood, and seeds was similar to that in the old nest, as was the "shape" of the nest. So upgrading nest size is unlikely to be the stimulus to move. Anyway, if size were the issue, the ants could simply enlarge the old nest, like adding an extra bedroom to a house.

Escaping ant guests seems like a weak reason too. Only the alleculid beetle larvae that steal food from incoming workers seem likely to be a bother at all. The spiders and reduviid bugs that prey on the ants are very mobile and do not live inside the nest. The abundant collembolans (springtails) on the seed stores eat mold (probably), and the spider (*Masoncus pogonophilus*) eats the collembolans. Crickets and silverfish are also scavengers, and hardly a problem for the ants. Moreover, it seems likely that many of these guests make their way to the new nest, either on their own steam or hitching a ride on seeds or ants. So this hypothesis falls flat, too.

This leaves contamination, disease, or parasites, but this hypothesis is untested. It is true that the floors of upper chambers in long-occupied nests are

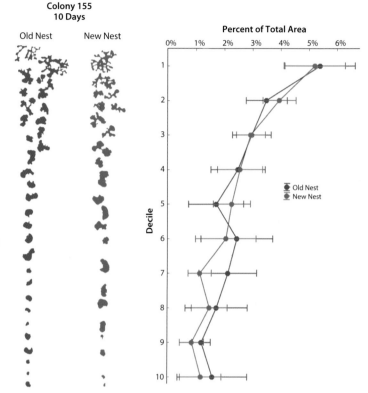

FIG. 5.3. The new nest the ants construct during a move is very much like the old one. *Left*, an example shows the chamber outlines of the old nest in blue, and of the new one (10 days later) in red. *Right*, averaged over multiple nests, the distribution of chamber area by depth (the nest's "shape") is similar for the old (blue) and new (red) nests. Modified from Tschinkel (2014).

often a black mat of (presumably) fungal filaments that sometimes bind the sand into a removable sheet, like a black linoleum floor. But there is no evidence that this mat is pathogenic or harmful in any way. It cannot be an aversion to black floors, for the ants cannot see in the dark. Still, until tested, some aspect of sanitation remains a viable hypothesis.

The world of ants is not our world, so there could be many reasons that we have not even thought of—and we are always sure there must be a reason because this conviction stimulates research. We can only test the hypotheses we have imagined, and thus we circle around the question. Eventually we find a reason that satisfies us, and then we stick to that story until there is compelling evidence to abandon it. And if the story is really good, it will be really hard to abandon. For the time being, we can suggest that there is some long-term advantage to moving often, but what that advantage is, we do not know.

The cost of moving

The energy, time, and risk of digging a new nest and moving into it should bear some kind of cost, and in order for the cost to be paid through natural selection, it has to be measurable in lost fitness. For an ant colony, this means producing fewer or less competent sexuals (winged female ants whose goal is to found a new colony, and winged male ants whose goal is to mate with these females), but this is difficult to measure, and we have not done so. The best we can do is use colony size as a proxy for fitness, because larger colonies, on average, produce more sexuals. But colony size is also measured indirectly using the disk area because this area is strongly correlated to the number of ants in the colony. If there is a fitness cost, then the disk areas of colonies that move more often should grow less or even lose more size compared to those that move less. Practically speaking, given an initial disk area, how much did it change over the two years of the study?

The complication here is that most colonies smaller than about 2,500 workers tend to grow, while larger ones tend to decrease in size. So the question has to be refined: Do colonies that move more often lose more size than those that move less often? The answer is yes, they do. Colonies that moved once or not at all lost none to 50% of their disk area, whereas those that moved two to four times lost about 80%. This loss of size suggests that these colonies were then less capable of producing sexuals—that is, their fitness was reduced. Of course, it could be that other correlates of moving were responsible for this size loss.

The progress of a move

A quantitative description of moves, in spite of being pretty simple, requires a considerable investment of time, more time than a single scientist could possibly have. My very patient employee Neal George had to sit next to a worker trail with a tally counter and make two-minute counts of the workers passing a point on the trail in each direction, as well as the number carrying seeds, brood, or charcoal. Moreover, he did this several times an hour from the time the nests became active at 9:00 a.m. until they shut down at 6:00 or 7:00 p.m. Then, in order to collect data at a higher rate, he visited three simultaneously moving nearby colonies over and over to count their trail activity. At the same time, he measured the soil surface temperature of the trail with an infrared thermometer.

So what did we learn? Most obviously, nest relocation is fast and well organized. Colonies just beginning to move were recognizable by the small disk of fresh sand around the nest entrance and a weak two-way trail connected to the parent nest. The move was initiated mostly by foragers. When we marked the workers digging this incipient new nest with fluorescent printers' ink, we found that after the move was complete, they were mostly foragers, the oldest age group of workers.

As the move proceeded, the trail traffic increased and the disk grew as excavation proceeded with vigor. In some cases, the workers cleared a conspicuous trail (fig. 5.4). Most moves were completed in four to six days, during which all showed a similar progression of trail activity—that is, they started slowly, reached a high intensity in the middle period, and leveled off as the move neared completion (fig. 5.5). When the move was finished, a few forlorn ants wandered around on the old nest like lost souls, but practically all their nestmates had moved on.

In keeping with the ants' strictly diurnal, workaday habits, trail activity and nest excavation began no earlier than a comfortable 8:30 a.m. and ceased by 6:00 or 7:00 p.m., when the workers closed both the old and new nests for the night. On hot days, trail traffic ceased for a few hours in midday when the surface temperature reached lethal levels of 45–50°C, sometimes even reaching 65°C. Heavy rain shut down the trail, but light rain did not.

The daily pattern of trail activity changed with the seasons. In June, a very clear, hot month, traffic was highest in early morning, low during the hottest midday period, and higher again in the late afternoon before it ceased for the night. The midday drop was less pronounced in cloudier July, and by the rains of August and cooler weather in October, morning activity dominated the moves.

Fig. 5.4. Workers sometimes cleared conspicuous trails during a move, but most moves took place without such route clearing. The old nest is the charcoal-covered disk at the top, and the new one is the disk of freshly excavated sand. Author's photo, from Tschinkel (2014).

A colony moves into a new nest gradually as the nest's size increases. It does this by means of two-way traffic on the trail, with a slightly higher rate of workers going toward the new nest than the reverse (fig. 5.5). Although the traffic can favor either direction for intervals, in our observations it averaged 12.4 workers per minute traveling to the new nest and 11.2 in the reverse direction. The small average excess of traffic to the new nest kept the new nest "topped up" as it grew, and the old nest gradually emptied. Summed over the whole move, workers made an average of 32,100 trips from the old to the new, and 28,400 trips from the new back to the old. The 3,800 extra trips from the old to the new nest during the move were in general accord with the average

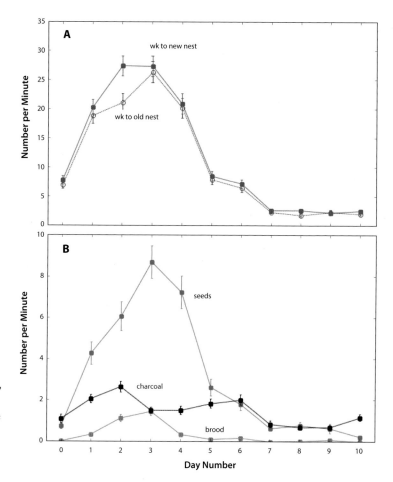

Fig. 5.5. *A*, the average worker traffic going to the new nest was almost always slightly higher than that of the reverse, so the colony gradually accumulated in the new nest. *B*, seed and brood transport generally paralleled worker traffic, but charcoal tended to be transported later in the move. Note the difference in the *y*-axis scales in these two graphs. Modified from Tschinkel (2014).

colony size at Ant Heaven. Since the average nest had only about 3,000–4,000 workers, it is clear that each worker made multiple trips.

Seeds, brood, and charcoal were always carried from the old to the new nest, almost never the reverse, and always at a much lower rate than worker traffic (fig. 5.5), showing that most workers made the trip unburdened. Brood transport estimates were especially low, probably because brood were transported in intense bursts that our sampling frequency often missed.

Why do the ants not keep it simple and have each worker make a single trip, more or less, from the old nest to the new, with each worker carrying something? Why this huge expenditure of effort, energy, and time, with almost 80% of workers making the trip empty handed, and several times at that? Why expose the workers to danger over and over? After all, workers carry seeds, brood, and charcoal only one way from the old nest to the new. Why do workers not make a single trip, or at least a small number? Instead, workers make 11 trips for every seed transported, 10 for every piece of charcoal, and 39 for every brood item.

What might be accomplished by the nearly equal two-way traffic? For one thing, it may simplify control of the emigration rate and eliminate the need for communication regarding the size of the new nest. Imagine that workers in the old nest make the trip to the new one only when they have been "informed" that there is adequate space there. This would require that some of the workers return after measuring the whole new nest and somehow pass this information on to workers waiting in the old nest. I suppose it is theoretically possible for a worker to take a momentary measure of the new nest by wandering all its chambers and shafts while counting steps, or some equivalent. Indeed, scouts of cavity-nesting ants and house-hunting honeybees take just such measures of empty candidate homes, but harvester ants would have to measure a relatively far larger nest and do so amid a beehive (sorry!) of activity, with workers running about carrying dirt and running into one another. It seems unlikely that they do this.

Now picture the alternative mechanism of continuous two-way traffic. A worker jogs to the new nest and enters it. Maybe she digs a little or carries out a bit of sand, but from the amount of jostling and crowding, she perceives that there is not yet adequate space and returns to the old nest. On the other hand, if she finds a bit of congenial space, she remains in the new nest. Only a small fraction of the workers entering the new nest every hour will find such congenial space and remain, creating the slight asymmetry in the trail traffic. As the nest grows to its full size, more and more workers will find congenial space and thus remain in the new nest, slowing trail traffic as a consequence. In this scheme, each individual worker decides whether to stay or not based on her own experience and does not need communication. The move from the old to the new nest is thus self-organized through the collective democratic votes of workers. Eventually most workers will find space in the new nest, leaving the old one more or less devoid of ants.

Even if we accept this scheme, it is still possible that those multiple trips are not spread evenly over the entire worker population, but that the number of trips varies among subgroups. For example, perhaps foragers make more trips than younger workers, some of whom may even make just a single trip from the old to the new nest. Similar considerations may apply to the transport of items, with some worker classes carrying more than others. Preliminary evidence suggests that worker classes participate differently in excavation of the new nest—many of the workers that we marked with fluorescent printers' ink as they dumped sand on the surface of the new nest turned up in the forager population after the move was complete, and they continued to enter the forager class for weeks, suggesting that these sand workers tended to be older than the average worker but younger than foragers. Many details remain to be worked out, but it is already clear that each move is a carefully orchestrated event, an ant ballet. Who plays each part is not yet entirely clear, but the choreography is there for us to see.

How do the ants know how deep they are?

A new nest is very similar to the old one in all aspects: size, chamber spacing, chamber shape, and vertical distribution of chamber space. This is also true of the vertical distribution of brood, workers of different ages, and seeds—brood reside in the nest depths with young workers, seeds in a narrow middle region, older workers in the upper regions, and foragers only in the top 20 cm of the nest. This vertical orderliness implies that the workers "know" what depth they are below the surface and thus excavate as well as arrange themselves, brood, and seeds in accordance with depth. This brings an interesting question to the foreground: What cues might tell them the depth? Such a cue would have to be very reliable and constant. This makes soil water and temperature unlikely cues, for both move downward in waves after rains and warming by the sun. Sometimes it is warmer and/or wetter above, and sometimes at depth. Although it seems a stretch, the cues could also be purely social if aging workers always moved higher in relation to younger workers. They would need cues as to the age of other workers and a sense of up and down (which they have, in spades).

Once we had more or less rejected soil moisture and temperature as reliable depth cues, it occurred to me that there should be a concentration gradient of CO_2 in the nest because the accumulated CO_2 given off by the metabolizing ants would vent to the atmosphere through the nest entrance, so concentration should increase smoothly with depth in the nest. If this turned out to be correct, a graph of this gradient would make a nice addition to my paper on the nest architecture of the Florida harvester ant, and it would suggest that the ants might use this gradient as depth information.

If I had a 2 m long syringe, I could sample the gas from chambers without disrupting the whole nest. Making such a syringe turned out to be unexpectedly easy, for our machine shop had stainless steel tubing 2 mm in diameter and 2 m long. I sharpened both ends, and there was my syringe.

It is always a good idea to do a pilot study before embarking on a project that will require a lot of effort or money, so I borrowed an ancient, unused infrared CO_2 meter from a plant biochemist. To collect the gas samples from the nests, I stoppered a dozen 500 ml Erlenmeyer flasks with rubber septa and pumped out the air with a vacuum pump. Now I had a dozen flasks with a vacuum that I could take out in the field. As I pushed my 2 m syringe needle down into the nest, each brief drop in resistance indicated that the bottom end of the needle had entered a chamber, at which point I pushed the top end through the septum of an evacuated flask, audibly sucking chamber air into the flask. It was just like getting your blood drawn—the phlebotomist pushes the free end of the needle into your vein and the other end through the septum of an evacuated test tube. Whereas you can see the blood squirt into the tube, you cannot see air flow into the flask, but a trust in physics assures you that it does.

Back at the laboratory, air from these flasks injected into the infrared meter showed that my hunch had been right—the concentration of CO_2 increased logarithmically the deeper the chamber, and it correlated strongly (negatively) with chamber size and worker age. If older workers preferred lower concentrations than younger ones, and if they dug more under lower concentrations, then the result would be both the distribution of workers by age and the relative size of chambers. Indeed, the curve showing the proportion of total chamber area by depth was the mirror image of the CO_2 gradient by depth. Moreover, insects can smell CO_2, and they respond to it behaviorally; they even have specific sensors on their antennae for that purpose, so this hypothesis seemed like a shoo-in. All that remained was to show that ants were actually smart enough to use this information.

I talked about this hypothesis with great assurance for several years before deciding that I needed to test it through experiments. I invested a good deal of money in a field-portable, flow-through, battery-powered CO_2 meter and repeated the sampling from live nests in the field. Once again, CO_2 concentration increased logarithmically with the depth of the chamber, confirming my earlier study. However, totally ant-free abandoned nests showed the same gradient, as did nearby soil with no sign of an ant nest. Clearly, the gradient was a property of the soil, with the ant nests in passive equilibrium with the surrounding soil. Once I thought about it, I realized that a few tired ants metabolizing in their nests would make hardly any contribution to the fired-up metabolism of billions of soil microorganisms. Had I been a soil scientist, I probably would have known that. On the other hand, it didn't matter how the gradient was generated; it could still provide depth information. Therefore, I needed to characterize how these gradients varied with site, time, soil type, and depth. To do this, I needed an easier way to withdraw gases from known depths at multiple sites throughout Ant Heaven and elsewhere. Drawing gas from non-cavity soil was slow and difficult, so I opted for a low-tech solution. A plastic vial with some holes punched in its lid and a length of fine surgical tubing glued through its bottom was buried upside down at several specified depths (15, 35, 75, and 175 cm) in holes bored with my new soil auger (I even splurged for the ratchet handle). The free end of the plastic tubing projected above the ground surface, where a syringe needle was inserted into the tubing (fig. 5.6). By con-

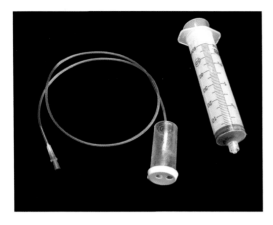

FIG. 5.6. The equipment for sampling gas from soil at various depths. The vial was buried upside down at a specified depth, with the tubing bearing the syringe needle projecting to the surface. Air in the vial equilibrated with neighboring soil. With the syringe connected to the needle, a soil gas sample could be drawn and analyzed. Author's photo, from Tschinkel (2013a).

necting a syringe to the needle, I could draw gas from the buried vial and inject it directly into the CO_2 meter.

Several sample arrays at Ant Heaven and several other locations in the flatwoods showed a lot of variation in the steepness of the gradient. When the water table was closer to the surface, as for example near wetlands, the gradients were much steeper, and CO_2 concentration just above the water table sometimes reached 4% to 5%. But at Ant Heaven, the water table is very deep (more than 5 m in some areas), so the gradients were more gradual, with CO_2 concentrations at 2 m depth 10–20 times ambient (0.3% to 0.8%), but varying twofold among the multiple sites. These concentrations offered reliable information about relative depth at all locations.

However, showing the universality of the CO_2 gradient did not prove that the ants actually used this information as a depth cue. This would require an experiment in which the gradient was removed from one set of experimental colonies and left in another. Could the gradient be removed? The sandy soils at Ant Heaven are very porous, so I expected that their included gases would diffuse rapidly. Indeed, the existing gradients were themselves evidence that this was so—the venting of the microbe-created CO_2 to the atmosphere at the ground surface was what created the gradient. Thus, if I could increase the area of interfaces with the atmosphere, the CO_2 would be vented across those surfaces, too.

FIG. 5.7. The setup for venting soil to the atmosphere and eliminating the CO_2 gradient. The black PVC pipes act as chimneys, drawing gas from the bottom of the 2 m deep boreholes. The sampling arrays are in the center of the borehole circle. Author's photo, from Tschinkel (2013a).

This was not very hard, given my fancy ratcheting soil corer. I buried a set of sampling capsules at several depths up to about 2 m and then bored seven or eight 2 m deep holes with this sampling array at the center (fig. 5.7). To speed venting, each hole was provided with a 3 m long PVC pipe with lateral holes near its bottom end. The pipes projected through screens to keep critters from falling into the holes, and their projecting parts were painted black so they would heat in the sun and draw air in from the bottom through the chimney effect. Sampling the central array for CO_2 showed that within a day or so the gradient had disappeared, and CO_2 concentration was no longer related to depth (fig. 5.8, *left*).

We were now ready for the next step—getting the ants to dig a nest in such a gradientless situation—but this too was not hard. We simply waited for a colony to start moving, and while the new nest was still very small, we surrounded it with a half dozen boreholes with vent pipes. Within less than a day, the ants were digging in gradientless soil, robbed of the potential depth information the gradient might have given them. We repeated this five times for vented colonies and five times for moved but unvented colonies (controls). What would they dig, and how would they arrange themselves, their seeds, and their brood?

To find out, we excavated, mapped, and censused each of the 10 nests chamber by chamber. Sadly for our CO_2 hypothesis, there was no difference in any

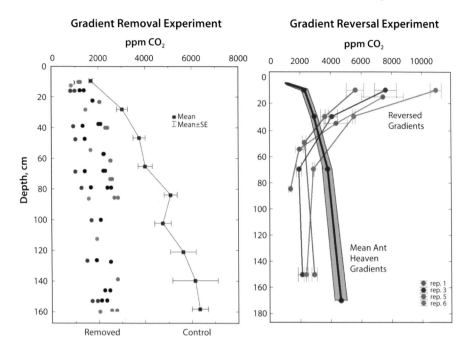

Fig. 5.8. *Left*, venting by means of boreholes eliminated the CO_2 gradient in five nests (blue symbols) in comparison with unvented control nests (brown symbols and line). *Right*, adding CO_2 to the top reversed the gradient (see text). Modified from Tschinkel (2013a).

measure between the moved-only and the moved-and-vented nests. The relative vertical distribution of chamber area ("nest shape") and the distribution of dark and callow workers, larvae, pupae, and seeds were essentially the same. The ants simply did not notice that the gradient they were supposed to use was not there.

This looked bad for the gradient-as-depth-cue hypothesis, but what if the test was not good enough, or too late, or what if the ants could detect very slight gradients? The correlation between the CO_2 gradient and these measures was so tight that there simply had to be something to it. I was facing the classic scientists' truth that correlation is not proof of causation, and my first experiment had suggested that this impressive correlation was not causal.

Great theories die hard

Hope for a terrific and appealing hypothesis does not die easily, and I decided I needed a stronger test. What if I could reverse the gradient and get colonies to dig nests in these reversed-gradient soils? To do this, I decided to eliminate the original gradient with six boreholes in a circle, as before, and then pump in CO_2 at the surface so it would diffuse downward in the soil. Toting a large tank of CO_2 to the field seemed to lack pioneering spirit, and besides, the tank might get stolen or someone might shoot a hole in it (after all, this *is* Florida). On the other hand, a visit to Florida Rock Co. bagged me a bucketful of crushed limestone (calcium carbonate). Dripping hydrochloric acid onto this limestone generated CO_2, the amount of which could be regulated by regulating the acid drip rate. I rigged a bottle of acid to a tree, with plastic tubing and a pinch-cock to regulate the drip rate onto the bottle of limestone below. The CO_2 thus generated was conducted to the experimental site with plastic tubing and dispensed from a porous tube under a plastic sheet that prevented direct loss of gas to the atmosphere (fig. 5.9).

A test of this apparatus showed that within a day or so, the CO_2 gradient in the central column of soil was completely reversed (fig. 5.8, *right*), with con-centration just under the surface 5 to 10 times that at depth. By planting a col-ony in this reversed-gradient situation, I could see whether the ants cued their digging and arranging to this gradient. Unlike in the first experiment, these colonies did not volunteer to dig in my prepared soil. Rather, they were exca-vated elsewhere and imprisoned in a screen-bottomed cage in which the only place to dig a nest was through a hole in the center of the screen, which they all did. If the ants used the gradient as a cue, then the smallest chambers should have been at the top of the nest and the largest at the bottom. The experiment was replicated four times, and the outcome was compared to normal nests dug in a normal soil CO_2 gradient.

Again, sadly, there was no difference in any measure between the control and the reversed-gradient nests. Both showed the normal distribution of cham-ber area in relation to depth. I finally had to admit that as appealing as my

FIG. 5.9. Setup for reversing the CO_2 gradient. *A*, the core of soil vented by means of boreholes and "chimney" pipes. *B*, hydrochloric acid dripping into a bottle of crushed limestone to generate carbon dioxide. *C*, a harvester ant colony was released into a screen-bottomed cage on top of the vented soil core. The ants accessed soil through a central hole in the screen. *D*, porous tubes that introduced CO_2 from the limestone reactor to the top of the vented core. Author's photos, from Tschinkel (2013a).

hypothesis was, the ants simply did not use these gradients as a source of depth information, at least not for the measures I had taken. So the final score was ants 100, Walter 0. The mystery of the depth cue used by the ants was still a mystery, and I would have to look elsewhere for how the vertical patterns arose.

Another hypothesis

With the demise of the CO_2 gradient hypothesis, what other mechanism might account for the nest architecture of the Florida harvester ant? The exponential decrease in chamber volume with depth (figs. 2.2, 3.11, 5.3) hints at some kind of probabilistic process. It might work something like this—on a worker's downward travel in the growing nest in search of a place to dig, the probability that she peels off from the crowd to dig is constant, so a fixed proportion of the still-downward-traveling group leaves to dig in each depth increment. As a result, the number of digging workers decreases with depth even if the workers don't "know" what their depth is. If there is, say, a 10% chance of a worker stopping to dig and there are 1,000 workers initially, then 100 will dig in the top nest increment, 90 (10% of 900) in the next increment, 81 in the next, 73 in the next, and so on. Fewer and fewer workers will thus dig at ever greater depth, and the volume they excavate will be directly proportional to their number. The total effect will thus be that the largest volume of soil is removed in the upper regions, with less and less at increasing depth, producing a fair facsimile of the general architecture of a harvester ant nest. There could also be an effect of worker age—young workers prefer greater depth and dig less (a known fact), while older workers prefer shallower regions and dig more (also known), with the net result of exaggerating the volume-depth profile.

An earlier experiment produced results compatible with such a mechanism. When I penned 200 workers from the top, middle, and bottom of harvester ant nests in cages, top workers produced the largest nests, middle the next largest, and bottom the smallest. The number of workers recaptured on the surface was in the same order, suggesting that the "digginess" of workers increases as they age and move up in the nest. On the other hand, there was no obvious difference in the "shape" of the nests—chamber volume decreased with depth in all of them at about the same rate. All were compatible with a probabilistic mechanism of nest excavation.

Was this another red herring, like the CO_2 gradient? Only experimental testing could tell, as we were once again faced with plausible explanations based on correlations. A test of this hypothesis was to force captive workers to travel down a copper tube to a depth of 40 cm before contacting soil that they could dig in ("40 cm access"). In the absence of depth information, the chamber size and distribution they dug at 40 cm depth should be similar to that in the uppermost region of a control nest ("2 cm access") because the number of workers digging there would not yet have been diminished by peeling off during their downward travel. That is, the highest rate of excavation would occur at the end of the pipe where the ants could first dig, and the nest should look like the control nest, but displaced downward by 40 cm. On the other hand, if the ants did have information on depth and applied it to chamber construction, then they should dig fewer and smaller chambers spaced farther apart. For simplicity of setup and interpretation, I created straight, chamberless shafts, angled downward at 50°. I had shown in previous experiments that the ants accepted a straight shaft as happily as a normal helical shaft and would festoon straight "starter nests" with chambers in the normal way.

The outcome of this experiment suggested that the ants have depth information and apply it to their digging efforts (fig. 5.10). The treatments in which the first soil access was 40 cm belowground dug fewer, smaller, and more widely spaced chambers, much as they would have in a control nest at that depth. Total chamber area in the 2 cm access treatment was over four times that in the 40 cm access, and the chamber area per worker was 2.5 times greater. This estimate

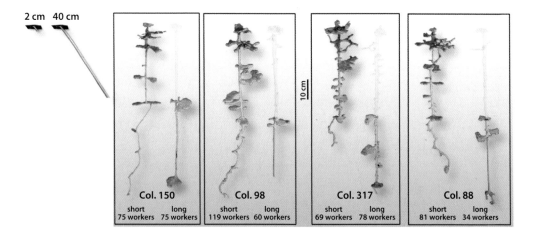

FIG. 5.10. When 200 workers first contacted soil at either 2 cm depth or 40 cm depth, they created chambers of a size and distribution similar to those in normal nests at that depth. This suggests that workers do have depth information and apply it to the chambers they dig. The number of workers accounted for after 10 days was about 70 for the treatment and 80 for the controls. The 40 cm copper tube of the deep treatment group is shown faded. Author's photo.

did not include chambers dug immediately under the screen, which, being open to the sky upon removal of the screen, were not part of the cast. In other words, both groups dug chambers similar to what they would have dug in a normal nest at that depth. Although this suggests that the ants know how deep they are, this experiment confounds depth with travel distance—rather than responding to depth itself, they are estimating the distance they have traveled downward belowground. Perhaps they count steps, as has been shown for foragers of some desert ants.

We realized we could test this depth versus travel distance hypothesis by simply installing the 40 cm copper tube at a shallow angle so its lower opening was only 4 or 5 cm below the soil surface, but the ants had to walk 40 cm to get to this point. The control was again a 4 cm tube that gave soil access at the same depth (i.e., 4 cm), but at 10% of the travel distance. As in the previous experiment, the copper tube led to a 40 cm deep, straight, chamberless, 50° shaft in both treatments. In a first run of this experiment, the ants escaped their Alcatraz by digging upward from the 4 cm deep junction of the shallow, 40 cm tube with the shaft. To improve prison security in the next run, I covered that junction with a large patch of screen so the ants had no choice but to get to work in the shaft below. This time there were no escapes, and after 10 days I made aluminum casts of what the ants had done (fig. 5.11).

The distant-travel group made complex chambers where they first had access to soil, and because these were under a screen buried 4 cm deep, these chambers were part of the cast. As in the first experiment, in the short-travel group, these under-screen chambers did not cast because removal of the cage removed their ceilings. Therefore figure 5.11 shows the under-screen chamber outlines for the distant-travel group, but those for the short-travel group are simply typical outlines, not from the test group. The fact that both groups created these complex chambers at very shallow depth suggests that they "knew" they were just below the surface. If we consider only this second experiment, we might think that the ants make these complex chambers wherever they first contact soil, but the two experiments together suggest that this is not true, for when first contact was 40 cm deep, the ants did not make complex chambers.

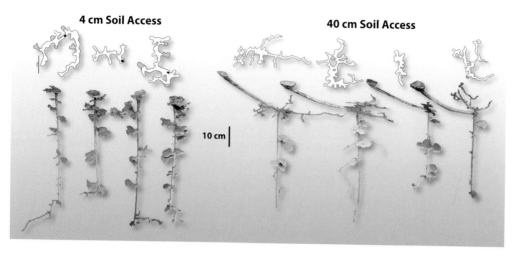

FIG. 5.11. In this experiment, soil was accessible at 4 cm depth in both groups, but at a travel distance of either 4 cm (*left*) or 40 cm (*right*). The chambers under the screen in the left group were open upon removing the screen and did not cast, but three representative examples of under-screen chambers from other colonies are shown. For the 40 cm access group, the uppermost chambers were cast because they were under a screen buried at 4 cm depth and were not open. Author's photos.

Figure 5.11 gives the impression that the total chamber area in the distant-travel group is less than in the short-travel group, but this is not a statistically significant difference, even if we leave out the top chambers. On the other hand, the distant-travel group made fewer chambers, much as they did in the deep-access group in the first experiment. Moreover, their deepest chambers were shallower than those of the short-travel group. They appeared to "know" when they were near the surface and used this as a guide to regulate chamber type in the upper regions (more tests needed), but they used travel distance as a guide for adding chambers. Once again, the ants have not given us a clean, unambiguous answer. Is it that they know their depth only when near the surface and respond to travel distance as a guide at greater depth? Do they have depth information for some functions (e.g. brood, seeds), but not for others (chamber number)? Such are the challenges of experimentation—there are always more questions.

My studies of nest relocation and construction, like most biological studies, illustrate the temptations of assigning cause and effect to correlation, and the difficulty of sorting causation from "mere" correlation. Our world fairly bristles with correlations that are incorrectly believed to be cause and effect. If you believe that flowering goldenrod causes hay fever in the fall, then you must also believe that hospitals kill people. After all, both are well correlated, goldenrod with hay fever and hospitals with death. Many a superstition has arisen from believing that correlated events are causally related. But a central drive in science is to create a system of causes and effects to explain the myriad complex patterns and correlations in the world around us. The key tool for doing this is the experiment—one varies the putative causal factor, measures the variation in the putative effect, and compares this to a control treatment in which the causal factor is unchanged. Sometimes very strong observed correlations, such as soil CO_2 gradients with nest architecture, turn out to be just that—correlations, not causes. Good experiments are deceptively hard to design and execute, largely because the exact nature of the control determines which putative causes the experiment actually tests. When I gave workers access to soil at 40 cm and 2 cm, the 40 cm treatment actually confounded (at least) depth with travel distance. When I vented the soil around the forming new nest, I surely changed more than just the CO_2 gradient. And of course there are always random differences among the replicates—chance differences in the composition of the worker groups, differences among the soils under each cage, differences in exposure to sun and shade, differences in the other creatures in the soil under each cage, including other ant species, and so on. These uncontrolled differences are the reason replicates are important, for they cancel out (more or less) these random sources of variation.

Few people appreciate how much effort, time, and often money go into establishing a single scientific fact of cause and effect. The science presented via TV, radio, magazines, or even books like this one usually provides a set of interesting facts and stories but rarely reveals how hard it was to establish each little point in these stories. Few people would have the required patience and tolerance of tedium or would be capable of deferring gratification for such long periods (sometimes years). But scientists have a great deal of curiosity and take pleasure in unraveling nature's secrets even if the promise of success lies far in the future, and even if the road is hard and tedious and paved with many failures.

Having established that ants *must* know how deep they are, we had to admit that we knew little about what information they used for this purpose, although we *did* know what information they *did not* use for this purpose. We can now turn to the equally difficult question of the "rules" ants use during nest excavation.

CHAPTER 6

A Diversity of Architectural Plans

As described in chapter 5, the nest of the Florida harvester ant is created when a colony relocates. The new nest is initiated by a small group of (probably) foragers joined over time by more and more workers from the old nest, most of which are not foragers. Because there are consistent and species-typical architectures, it is reasonable to ask whether the ants have an architectural plan—that is, a set of "rules of excavation" through which the nest comes into being. Through what mechanism do the many characteristics of the nest arise—overall size; chamber size, shape, and spacing; total volume and area; distribution by depth; and other attributes that we so easily recognize? For example, what rules create the lobed chambers and helical shafts of a harvester ant nest, and what rules create the smaller relative size and spacing with depth? Answering these questions is a large and important project, and for any hope of success, it must be divided into smaller questions that test one or a few attributes at a time.

NEST SIZE

One of the most obvious nest attributes is that large colonies dig large nests and small colonies dig small ones. How does the colony achieve such size-appropriate nests? Do the harvester ant workers digging the nest "know" how big to make it? If so, then workers from small colonies should build smaller nests than those from large colonies. I tested this by penning 400 workers from large and small colonies in screen-bottomed cages and allowing them to dig a new nest through a central hole in the screen. If these workers had any memory of the size of their home nest, the nests they dug would reflect that size, but they did not. Perhaps not surprisingly, they all dug nests of a size appropriate to the number of workers digging, not to the size of their home nest. This suggests that no "memory" guides the excavation. Information on size is not stored in a group of knowledgeable skilled builders, but in the whole population of workers.

Of course, this experiment is unrealistic in the sense that during nest construction for a move, the number of workers digging is not fixed at the outset but gradually increases. But it didn't matter how I added the workers in a second experiment, either 100 a day until there were 400, or 400 workers all at once at the beginning of the experiment. Once again, the two groups dug similar nests, suggesting that it is the *final* size of the group that determines the nest architecture, not the gradual increase in group size. A number of laboratory experiments have shown that some feedback from the nest size eventually slows and stops excavation at a size proportional to the number of workers. This is undoubtedly what happens during a colony move and is why the new nest is so similar in size to the old nest. But what this feedback might be is currently not known. It seems unlikely that it is simple crowding, for crowding differs so much from the top to the bottom of the nest. We can add it to the list of unsolved mysteries.

Digging a nest is work whose difficulty and rate must depend on the nature of the soil the ants are excavating. Harvester ants are creatures of sandy soil, so I took 400 workers from colonies and made half of them dig in their native sandy soil, and the other half in the more clay-rich soil of the Tallahassee Red Hills. Clay is much more resistant to digging efforts than sand, as I can personally attest. The nests dug in clay soil averaged 40% smaller than those in sand (fig. 6.1), suggesting that nest size was proportional to the effort and/or time required for excavation. The smaller nests were also less complex, but this might have been simply a function of their smaller size. In any case, this experiment established that the nature of the soil affects at least the size of the nest, and possibly its architecture as well. It seems likely that this applies to most ant species.

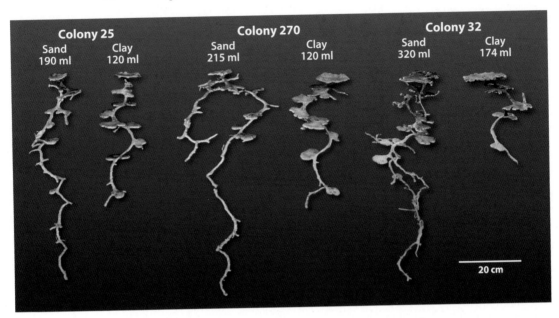

FIG. 6.1. Nests excavated in denser, more clay-rich soil are smaller and simpler than those in sandy soil. Each pair of nests were made by 200 workers from the same source colony, and thus the two nests in a pair are directly comparable. Total nest volume is given under the soil type for each colony. Author's photo.

CAN WE GET THE ANTS TO REVEAL THEIR ARCHITECTURAL PLAN?

Figuring out how the ants create their complex nests underground and in the dark is not easy. We can't see what the ants are doing, and even if we could, we would have to see what each individual ant is doing, how she interacts with other ants, and how she responds to cues emanating from the nest under construction. Most of my architecture experiments depend on penning harvester ant workers in order to make them do my bidding. This works because the ants are compulsive diggers. They just can't help themselves—give them sand, and they dig. If the sand is damp, they trowel it into pellets (fig. 4.7); if it is dry, they dig like dogs, pitching sand backward between their hind legs, their front legs a blur. They may pack the dry sand into their psammophore, a basketlike set of hairs under their head, and carry it out this way. Their digginess makes it possible for me to get answers in a few days. The challenge is to design and carry out experiments that give clear answers.

We can make the ants dig in sand between plates of glass (a "sand"-wich), which they happily do, and because we can see the ants through the glass, we can see which individuals do what, and how consistently they do it. This is certainly interesting for observing how digging is organized, but in my experience, what they produce under glass usually bears little resemblance to what they do in natural soil. In such sand-wiches, harvester ants dig snaking, branching tunnels and irregular chambers that somewhat resemble the complex uppermost chambers of natural nests, but usually not the simpler oval chambers found at greater depth. Frustratingly, for no obvious reason, they do sometimes produce nice simple chambers. I have been unable to find conditions under which the ants consistently dig chambers similar to those deeper in natural nests. In any case, the natural nest is composed of many chambers whose shape, size, and spacing change with depth. The complexity of this endeavor is thus far greater than digging a single chamber of any type.

But rather than despairing about the difficulty of this challenge, what if we could *ask* the harvester ants to share with us their "opinion" of what a natural harvester ant nest should look like? We could ask them this question if we could offer them complete nests of our own construction. We could vary features of these nests to deviate from what we have observed in natural nests through castings and excavations, and see how much these "Franken-nests" have to deviate from the natural before the ants respond by modifying them or rejecting them. Simple enough, right? For months I ruminated about how one could construct hollow spaces of one's own design underground, until one day it came to me—water can be frozen in any desired shape in a mold. When this ice is buried, it will melt to create a hollow space in the shape of the ice. I soldered some copper strips to copper bases in the shapes of a range of harvester ant chamber types and sizes (fig. 6.2, *top*). These I filled with water and froze. When they were unmolded I had facsimiles of nest chambers in ice, 1 cm thick just like real chambers (fig. 6.2, *bottom*).

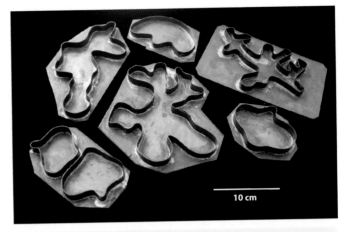

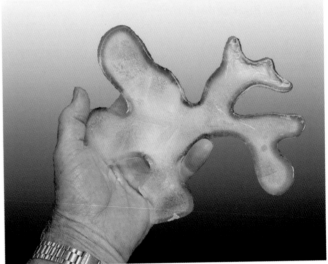

FIG. 6.2. Freezing water in copper molds (*top*) results in ice facsimiles of the subterranean nest chambers (*bottom*). Author's photos, from Tschinkel (2013b).

The next step was to take these to the field, dig a pit, place the lowermost chamber in the bottom with a plastic tube against its side, and quickly cover it with sand before it melted (fig. 6.3, *left*). Then pack more sand to reach the depth of the second deepest chamber, at which point the process was repeated until we had buried the uppermost chamber just below the surface. By pulling the plastic tube out, we created a shaft connecting all the ice pancakes, and when the ice melted, we had created subterranean spaces according to our plan. I was now able to make such an ice nest with any depth, spacing, chamber shape,

Fig. 6.3. *Left*, burying ice chambers, with a plastic tube connecting sequential chambers. When the tube is withdrawn and the ice melts, a facsimile of an underground nest is formed. Author's photo, from Tschinkel (2013b). *Right*, an aluminum cast shows that the artificial nest is very similar to a natural one. Author's photo.

and size that I wanted. Aluminum casts of these ice nests showed that I could make nests that were nearly indistinguishable (at least to me) from natural nests (fig. 6.3, *right*).

By offering these nest facsimiles to the ants, we could see whether and how they modified them, thereby telling us how they like our real estate. To make sure that the ants didn't wander off into the sunset, I placed an escape-proof screen-bottomed cage over our creation so the ants could get underground only through a hole in the screen that was positioned over the entrance to our test nest. I added the number of ants found in a natural nest of the same area. The ants entered within minutes, and after an hour or two they were busy bringing sand to the surface. After three to six days, I recaptured all the ants I could and returned them to their own nest, their service done, and then made an aluminum cast of the test nest. If the ants consistently modified or did not modify particular features built into the ice nest, we could paint a picture of what they expected a natural nest to look like. But we should be clear that the "opinion" we sought was not that of any individual ant, but a collective expectation that existed only at the colony, not the individual, level.

Chamber order

One of the most obvious features of natural nests is the change in chamber shape and size with depth, with the largest, anastomosing chambers uppermost and the smallest and simplest at the bottom. When we offered an ice nest of this construction, the ants modified it relatively little, changing the original area only a few percent (fig. 6.4, *left*). This suggests that they didn't find this very different from the nest they would have dug themselves. However, when we reversed this order so that the big, branching, coalescing chambers were at the bottom and the small, simple ones just under the surface, the ants registered a big complaint by digging a large, branching chamber similar to those in natural nests just under the surface, obliterating the tiny chamber we gave them (fig. 6.4, *right*). This increased the chamber area in the upper nest regions by more than 350% over the ice chamber area. At the same time, the ants also partially filled the large chambers in the middle and bottom regions, decreasing their area by 30% to 40%. Less than 5% of the area was filled in the upper regions of the normal-order nests. The workers asked to "judge" both the normal and reversed ice nests were all from the same source colony, so differences in the outcome did not reflect colony differences but were responses to the offered nest. The net effect was to make the reversed-order ice nests more like the normal-order nests.

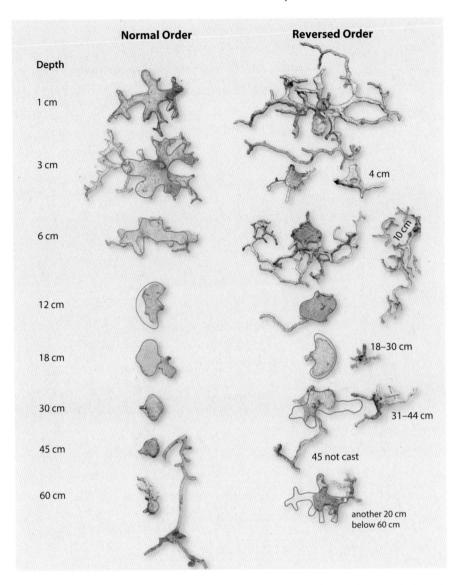

Normal Order **Reversed Order**

Depth

1 cm

3 cm

4 cm

6 cm

10 cm

12 cm

18 cm

18–30 cm

30 cm

31–44 cm

45 cm

45 not cast

60 cm

another 20 cm
below 60 cm

Fig. 6.4. Ants were offered ice nests with a normal order of chambers (*left*), or with the order (but not the spacing) reversed. The outlines of the ice chambers are shown in pink and are superimposed on images of a cast made after four days. Chamber depth is shown on the left. The ants greatly modified ice nests with a reversed chamber order. One of three replicates is shown here. Author's photo.

In the reversed nests, the ants did not stop at enlarging and filling ice chambers; they also added chambers where there had been none, increasing the total nest area by 20%. Eighty-five percent of this area was in the top 10 cm of the nest. In contrast, the ants in normal-order nests added no chambers beyond the offered ice chambers. These differences are obvious in the example in figure 6.4.

Chamber spacing

Another conspicuous and testable nest characteristic is the spacing between chambers. The topmost chamber is usually 1 or 2 cm belowground, and chambers following that are 1 to 3 cm apart. Spacing between the bottom chambers is normally severalfold greater than that between upper chambers. So we offered the ants an ice nest in which the normal top chamber was 10 cm rather than 1 cm belowground, the next 15 cm deeper, and so on until the (normal) bottom chambers were spaced only 1 to 3 cm apart. The comparison was again to ice nests with natural size, order, and spacing.

Ants in two of three nests created large, complex chambers just under the surface where there were no ice chambers, but overall the nests were not significantly different from the normal nests, perhaps because a normal, large chamber was available at only 10 cm depth. Nevertheless, the ants expected large, complex chambers just under the surface, and if they didn't find them, they tended to make them.

Horizontality of chambers

Another impressive and consistent feature of natural nests is the absolute horizontality of the chamber floors (fig. 4.11), and this too was testable. It seems a bit mean, but we offered the ants ice nests in which every chamber was tilted drunkenly about 20° from horizontal, but their sizes, shapes, order, and spacing were natural. Surprisingly, the ants seemed rather unconcerned that their floors were sloping and their furniture was sliding to the bottom. Perhaps if given more than a few days, they would have taken this situation in hand, but it was not their first impulse, and they accepted the tilted chambers as they were. We can surely expect that any new chambers they excavate will have horizontal floors, because that is what they do.

Shaft variations

Figure 4.10 shows how fond the ants are of making helical shafts. The ubiquity of this feature makes it a hallmark of harvester ant nest architecture and suggests that the ants would reject any other shaft configuration. We were able to readily test this by comparing how ants responded to a helical shaft (without chambers) versus a straight shaft of the same slope (also without chambers). We bent an aluminum rod into a helix, screwed it into the ground, and then unscrewed it to leave a helical shaft with a radius of about 5–7 cm and a slope of 53°. The straight shaft was made with a straight aluminum rod pushed into the soil at either 10°, 53° (natural slope), or 90°. The lower end of the 10° shaft was only 13 cm belowground, whereas in the 50° shaft it was about 30 cm, and in the 90° shaft, 50 cm belowground. One of our famous screen-bottomed cages was placed over each shaft entrance and 150 workers were added to each; they were then given four days to accept or reject the shafts, add chambers, or modify them in other ways.

Surprisingly, the ants seemed to mostly accept and use both straight and helical shafts (fig. 6.5). Helical shafts were always accepted and decorated with chambers. Straight shafts inclined at 53° were accepted in 13 of 14 cases—only once did the workers abandon the offered shaft and build a helix of their own. In contrast, when the inclination was only 10°, all four groups festooned this shallow shaft with two descending helical shafts. Similarly, one of four groups that were offered a 90° shaft abandoned it and made a helix of their own. The somewhat surprising conclusion was that the ants seemed more concerned with the angle of descent than the unrelenting curvature of a helix (fig. 6.5). On the other hand, when the ants reached the bottom of the straight shafts, they deepened the nest by means of their hallmark helical shafts.

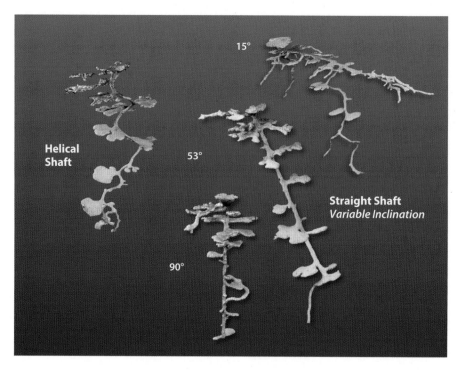

FIG. 6.5. Experiments in which ants were offered shafts without chambers. Ants universally accepted helical shafts, decorating them with chambers, but they also almost always accepted straight shafts when these had about the same angle of descent as the helix (53°). While they used straight shafts with a 10° angle of descent, they always added helical shafts. They also used straight vertical shafts in most cases. Author's photo.

Chamber placement

In spite of the relatively undiscriminating acceptance of shafts, the placement of chambers was very different on helical versus straight shafts (fig. 6.6). In all three helical shafts, practically all chambers (23 of 25) were on the outside of the helical curves, only 3 were on both sides, and none were on the inside. Like a meandering river, the ants have a strong tendency to cut into the outside of a bend. In contrast, chambers were about equally likely on both sides of the 14 nests with straight 53° shafts and were often even opposite one another (fig. 6.6). Of the 109 chambers on these straight shafts, 55 were on both

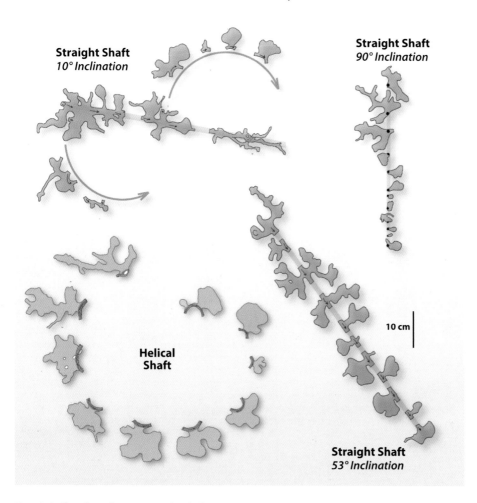

Fɪɢ. 6.6. Chamber placement in the shaft experiments. Chambers were almost always on the outside of the helical curve, but on either or both sides of the straight 53° shafts. With a vertical shaft, the chamber was always to one side of the shaft, but with random compass orientation. The 10° shafts were widened to both sides, often with indistinct chamber formation, but always had ant-made helical shafts descending from them. Author's photo.

sides, 31 were on the right side, and 23 were on the left side—not much of a preference. With a 10° slope, the shafts were usually widened to both sides, often without distinct chambers, and in the 90° shafts, the shafts never entered the middle of chambers but were always to one side of the shaft with random compass orientation.

MISCELLANEOUS QUESTIONS

Sometimes, the ants come up with original ways of attaching doubt to the results. For example, I have several times asked the ants whether they care about chamber shape, or whether any shape chamber, as long as it is large enough, will do. To do this, I presented them with ice nests in which the chambers were circular or triangular (with the vertex at the shaft). In some cases they accepted both of these chambers. I know they did because there were seed stores in them. In others, they partly filled the chambers with sand. But the response that really adds doubt is that they often decorated the ice nest with extraneous branched, "wild" shafts. Does this mean that given enough chamber space, they put their energy into making shafts? I am faced with trying to decide what to believe—do I decide that the circular and triangular chambers are OK with them, or do I decide that the wild shafts are a "protest" against the strange chambers I offered? Such are the difficulties of getting into the mind of a colony.

In natural nests as well as in my "shafts experiment," incipient future chambers are visible as little bumps or angles on the outside of the helix and are spaced much like chambers (fig. 6.7). By what mechanism these little bumps arise is a significant question, for they determine the eventual spacing between chambers, acting as a sort of placeholder or road sign—dig here! Are they the sites of regular traffic jams, with ants making little deviations to get by? And why do some grow into chambers and others not?

DEPTH CUES REVISITED

To create a vertically organized nest, the ants must have information on depth below the surface. We already know this information is not derived from the CO_2 gradient (chapter 5). What if the ants "know" their depth through some cue embodied in the soil itself? A desperate, long-shot hypothesis for sure. Let us get a set of workers in a screened cage to dig a nest starting from the surface and another set starting from the bottom of a 50 cm pit. If the ants sense that the soil in the bottom of the pit is "50 cm soil," they should create much simpler, smaller chambers at the top of their nest than the ants digging from

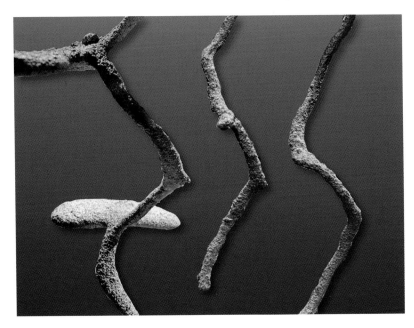

Fig. 6.7. Chambers are usually formed at slight angles or bumps on the helical shaft. This is particularly apparent in nests being actively excavated but is often also visible in the deeper shafts of more stable nests. The spacing of these bumps and angles is similar to chamber spacing and probably indicates where a chamber will be formed as the nest expands. Author's photo.

the surface. In fact, they should dig chambers similar to those normally found at that depth. Sadly, in spite of the great deal of work and sweat required to set up this experiment, there was no difference in the nests the ants excavated, suggesting that the soil contains no "intrinsic" depth information that the ants use. Sometimes we try something that seems logical, but the ants have a different opinion. We learn from these negative results, too.

My efforts to figure out how harvester ant colonies create their elaborate nests barely scratch the surface. The basic mystery remains mostly intact. It remains for others to make further inroads. In the next section, I turn to another question that cries for an answer.

HOW DOES NEST ARCHITECTURE SERVE THE COLONY?

Generally, the nest serves as shelter, protection, and microhabitat. However, it is much more interesting to consider whether and how the "design" of specific elements of nest architecture serve specific colony functions better than other designs, in the sense that they help maximize reproductive success. The enormous variability of architecture across species suggests that this might be so. A specific architecture may be tailored by evolution to best serve the colony that built it. Sometimes the purpose is fairly obvious—the egg-shaped chambers of fungus-gardening ants are clearly more suited for holding fungus gardens than would be the pancake chambers of most other species. On the other hand, it is not obvious how chambers with elongated lobes might be advantageous over those with simple oval outlines, or why spacing similar chambers closer or farther apart serves each colony better than the alternative.

But is the belief that each particular architecture is a *specific* adaptation just an application of the belief that everything in biology must be the optimal product of natural selection? As much as this belief (often referred to as "adaptationist") stimulates scientific inquiry, it is not necessarily always true, or at least supporting evidence may be difficult to impossible to accrue. In the previous chapter, I showed that harvester ants are very definite about some rules of construction and have no strong opinion about others. They insist on complex, branching chambers just under the surface, as well as chambers on the outside of curved, helical shafts, but if the shafts are straight, they have no preference for sidedness. They accept the offered nonhelical shafts, but on their own, they make a helix. We learned that they prefer but are not fanatic about a steep downward angle of shafts and are unhappy with very shallow angles of shaft descent. They often accept a vertical shaft but often reject it as well, suggesting a sort of "can we live with this" moment. These experimental results tell us a few things about *how* the ants create the lovely nests they create, but they tell us almost nothing about *why* they create nests with such very particular features. It is one thing to nibble around the edges of the rules that govern nest construction, but it is an entirely different and more difficult thing to figure out how the results serve colony fitness.

To answer the *why* question, we must determine the effect that *particular* architectural features, separately and together, have on colony fitness—that is, on the success of a colony in producing daughter colonies. Such experiments are extremely difficult, not so much in theory as in practice. Let's say we want to ask whether one of the conspicuous features of a natural nest, for example the helical shaft, affects colony fitness. As a reasonable proxy for colony fitness, we can compare the number and quality of sexual ants a colony produces in a nest with a helical (natural) shaft versus one in which the shaft is straight with the same angle of descent. We must force colonies to use these nests for an entire annual cycle without modifying the nests, so that we can contrast the sexuals produced in these modified nests with those produced in a nest with natural architecture. Similar conditions would apply to modifying other features such as chambers, total area, nest depth, distribution of chamber size and depth, and so on. Some of this could probably be done in the laboratory, but here, the natural rates of mortality and the cost of foraging would be absent, with unknown effects on our measures of fitness. But even in the laboratory, colonies are not easily kept healthy and happy for so long. By any measure, these experiments are daunting to impossible.

A TEST OF AN ARCHITECTURAL FEATURE

Nevertheless, some features are probably more amenable to testing than others, especially if we are satisfied with shorter experiments. I completed such a test of an architectural feature, but on a far less finicky ant than the Florida harvester ant, namely the exotic fire ant, *Solenopsis invicta*. In 1964, the famous bee biologist Charles Michener published a paper that described how he sampled the makeup of the colonies of a number of bees, wasps, and ants and suggested that the larger the colony, the less efficient it was at rearing offspring. That is, the number of offspring per adult decreased as colonies grew larger. Michener wondered why these insects had evolved sociality if it meant that their reproductive success decreased as a consequence. This eventually became known as the "Michener Paradox." However, instead of seeing this question as a hypothesis to be tested, entomologists, including me, accepted these results down through the decades as a kind of "true fact." Indeed, Sanford Porter showed as part of his PhD work in my lab that colonies of 3,000 workers reared fewer new workers per worker per month than did colonies of 750 workers. This result seemed in accord with Michener's report.

In the early 1990s, I excavated and censused 90 (90!) fire ant colonies over a one-year period. These censuses revealed the number of workers, larvae, pupae, and sexuals for a wide range of colony sizes across the seasons. The birth rate of new ants was estimated from the number of workers and pupae, adjusted for the ambient nest temperature. It works like this: If the pupal development period is, for example, 10 days, and there are 1,000 pupae, then the raw birth rate is 100 per day. If there are 1,000 workers in this nest, then the specific birth rate is 0.1 new worker per worker per day. By making this calculation for natural fire ant colonies ranging from a few hundred workers to a quarter million, I found that this birth rate (or production rate, or brood-rearing efficiency) did not change with colony size, although it did change with the season.

For years, I puzzled over this contradiction between my field results and the earlier lab results, as well as the Michener paper. We tended to look for differences that might explain this contradiction, and early on, it seemed to me that the answer might lie in the fact that whereas the lab colonies with their thousands of workers had been reared in a single large chamber, in nature the colony occupies a nest composed of hundreds of small chambers, each with space for 200 ants, on average. Since we had already shown that the smaller the lab colony, the higher the efficiency or birth rate, it made sense to ask whether natural colonies maintain constant efficiency because the colony is divided into hundreds of small, efficient "working groups." Because the average chamber size does not change with colony size, this might explain why large colonies are as efficient as small ones.

This was clearly a testable hypothesis about how nest architecture may affect colony function. All I had to do (easier said than done) was rear equal-sized colonies in two types of laboratory nest—either a single large chamber, or 48 small connected chambers of the same total area as the large chamber. Over the years, I had two or three undergraduates attempt this experiment, but the results, if any, were equivocal. So, as a final act under one of my final grants, my assistant Nicholas Hanley and I set the experiment up and ran it twice.

I used a router to carve large and small chambers in flat blocks of plaster, establishing once again that one should not rout dry plaster with a bit spinning at 25,000 rpm, as my entire shop had a white coating after only a handful of nests. The finished nests were placed in trays, dampened, covered with glass, and then stocked with 3,000 workers and 3,000 brood (fig. 6.8). After feeding ad libitum for one brood cycle (about a month), each nest was opened and censused to establish how much brood the original workers had reared. The treatment was then reversed, so that colonies from single-chamber nests were now housed in multiple-chamber nests and vice versa, and the whole experiment was run for another brood cycle, with a final census at the end.

As much as I would like to have found a positive result—that is, that divided nests were more efficient at rearing brood—neither run of the experiment gave me that satisfaction. Both showed that brood-rearing efficiency, no matter how measured, was similar in divided and single-chamber nests. Nest

FIG. 6.8. The experimental setup for testing the effect of nest subdivision on brood-rearing efficiency. The two nest types had the same total chamber area and started with the same number of workers and brood, along with a single queen. Author's photo, from Tschinkel (2017c).

subdivision could not resolve the difference between our laboratory results and the field results.

At about the same time, my colleague Bob Jeanne at the University of Wisconsin joined two other colleagues to reanalyze the data on which Michener had based his 1964 paper. They were motivated by several empirical studies published since 1964, including my field study of fire ants, that had found no decrease in brood-rearing efficiency as colonies grew. Jeanne and his colleagues showed that if they considered aspects of life cycles, seasonally varying mortality, and a number of other natural history factors, the so-called Michener Paradox disappeared. So I had done my experiment in vain, testing a hypothesis that turned out to be fictitious. In retrospect, I should have seen a strong warning in the work of my graduate student Deby Cassill, who showed that the brood piles of fire ants were always covered with workers. In other words, the larvae got the same care even if there were far more or fewer workers than larvae, or whether the larvae were nestled in small crèches or huge feedlots. The efficiency of brood rearing depended on the ratio of brood-care workers to larvae needing care, and this ratio was always the same.

This story is a cautionary tale about the dangers of accepting prevailing beliefs about a published paper, rather than reading the paper with a critical eye before embarking on a project stimulated by that paper. From the nest architecture point of view, this outcome was a bit saddening, for the nest division was one of the few experiments on the function of nest architecture that I could actually carry out. Some of the conditions nests provide, such as humidity or moisture, affect brood rearing and other colony functions, but they are not themselves the architecture. Until someone thinks of a way to directly test the cause-and-effect functions of the architecture itself, we will have to be satisfied with patterns and correlations (of which there are many).

Failure to explain the function of architectural features does not close other avenues of inquiry. In the next chapter, I will take up the topic of the important effects the excavation of nests has on the soils in which the ants live.

CHAPTER 7

Ants and Soils

A whole unfamiliar world lies belowground, a world that demands many special adaptations because, in contrast to the aboveground world, the medium is dense and granular. Yet the world's soils are home to a diverse and abundant flora and fauna, many of which shuttle between above- and belowground, some resident in both, and others never emerging above this boundary. While natural processes relentlessly move dissolved minerals downward and create distinct zones (horizons) in the soil, animals that burrow in the soil slow these nutrient losses and redistribute soluble minerals by bringing excavated soil to the surface to form topsoil (biomantling), or perhaps just to higher levels belowground (bioturbation). Soil structure is thus constantly altered and renewed by these creatures as they go about their lives. Their actions make nutrients more available to plants, affect the infiltration of water, increase aeration, and change soil porosity. Over the millennia, many soils are like a slowly boiling pot of viscous liquid, with plumes rising and spreading as they breach the surface, or falling to create turbulent mixing.

One of the effects of this stirring of soil is that objects on the surface gradually sink belowground and, if they are not subject to decay, are preserved for very long periods. When my brother lived in Carthage, Tunisia, simply digging a hole to plant a posy would often turn up a Roman coin or two. Archaeologists depend on this burial phenomenon for their livelihood, and gardeners curse it because the paving stones in their neat walks gradually disappear belowground. Charles Darwin published one of the first studies of this burial process and attributed most of it to earthworms. Some contemporaries of Darwin considered this a relatively trivial topic for such a broad conceptual thinker as Darwin, but of course, like evolution by natural selection, it was an example of large effects through small actions over a long time.

Darwin did not assign a large earthmoving role to ants, perhaps because England's ant fauna is rather anemic. However, it did not take long for others to recognize that in most of the world's warmer regions, ants are major agents of burial and bioturbation. Most of this action is the result of the excavation of subterranean nests. The rate at which such deposits bury the soil surface obviously depends on the volume of the nests and excavated soil as well as the density of the nests. The enormous and expansive nests of tropical American leaf-cutter ants bury the landscape under many cubic meters of soil, and no one who views their extensive piles and nest openings could fail to appreciate that these ants are major agents of bioturbation. However, nests do not have to be large and obvious for the burial rate to be significant. A high density of ants with small nests can bury a garden patio at a remarkable rate.

SOIL DYNAMICS AT ANT HEAVEN

No one has yet characterized the effects of an entire ant fauna on the stirring and burying of soil in any habitat. I did most of my own work on this subject mostly at Ant Heaven, where, in addition to the Florida harvester ant, *Pogonomyrmex badius*, there are dozens of ground-nesting ant species. As each colony excavates its nest, it brings increasing amounts of soil to the surface as it grows and probably deposits varying amounts belowground as well. If, like *P. badius*, these colonies also move to a new location now and then, the amount of soil collectively turned over increases in proportion to the frequency of moves, nest density, and the volume and depth of each species' nest. At Ant Heaven, as in many ecosystems in the warmer world, these effects on soil dynamics are probably large. Together, the ants (and other burrowing animals) slowly stir the soil on a multimillennial timescale, bringing soil to the surface only to bury it again, reexcavate it, place it belowground, bring it back to the surface, and so on ad infinitum. Of course ants are not the only creatures that affect soil dynamics, but they are probably the most important.

My adventures with soil dynamics began when an archaeologist, Jim Dunbar, and a sand chronogeologist, Jack Rink, contacted me. They had seen my work on ant nest architecture and speculated that I might be able to help them with a conundrum. During an archaeological dig at Wakulla Springs south of Tallahassee, they had found flint chips—reliable evidence of Paleo-Indians—about 1 m deep. Normally, such chips would be a merry discovery, but there was a problem—dating the host soil by optically stimulated luminescence (OSL) resulted in a date between 30,000 and 40,000 years ago, long before all other evidence suggested the presence of humans in the New World. OSL is based on the fact that ambient radiation gradually causes more and more high-energy electrons to be trapped in quartz grains the longer they are kept in the dark, for instance if they are buried. A brief flash of light in the laboratory causes this trapped energy to be emitted as ultraviolet light. Normally the deeper the grains are buried, the longer they have been buried and the more ultraviolet light they emit. Dunbar and Rink suspected that bioturbation had scrambled the soil so that older, deeper soil was mixed with younger, shallower soil, resulting in falsely old dates. The chief suspects for this skulduggery were ants, and this is where I came in. Could we perhaps collaborate on experiments to quantify the amount of bioturbation and biomantling of which ants are guilty in our area?

The question was whether harvester ants move a significant amount of soil from deeper to higher levels but still deposit it belowground, thus without exposing it to light. Such immigrant sand would change the capacity of the host layer to emit light under OSL and would thus result in false dates. Knowing how much of this went on might help archaeologists adjust for the action of ants and correct OSL dates. Soil deposited on the surface was of little interest, as exposure to light immediately bleached its ability to emit ultraviolet light in OSL, or as Jack phrased it, zeroed it out.

Jack ordered 2,000 pounds of sand of 11 different colors and had it delivered by truck from Pensacola to Ant Heaven. Who knows what the vendor thought we were doing having sand delivered to the middle of the woods in an area that was already pure sand? We dug two pits, each 2 m deep and measuring

1×2 m in horizontal aspect, piling the excavated sand on tarps. For a good deal of the excavation (and in my absence), three grown men and one woman watched while Christina, who measures 157 cm, pitched sand out of a pit that was deeper than she was tall, and that she couldn't get out of without help.

The pit finished and squared, we lined three walls with plywood to form a 1×1 m space 2 m deep and filled this with layer after layer of different colored sand, each layer 10 cm thick, with the bottom meter of two layers, each 50 cm thick (fig. 7.1). Each color was diluted with three parts natural sand using a rented cement mixer. The adjacent 1×1×2 m pit was the working space and was filled with natural sand as we piled colored layer upon layer. We now had two layer cakes for which we knew the depth of each color of sand.

Christina and I excavated a large colony of harvester ants and planted it with a couple of thousand workers every two days, queen last, in a screen-bottomed cage on top of the layer cake. The colony was thus confined and had to excavate a nest through the hole in the middle of the screen bottom. Because the

Native
0–10 cm

Chocolate
10–20 cm

Green
20–30 cm

Violet
30–40 cm

Blue
40–50 cm

Pink
50–60 cm

Black
60–70 cm

Lime
70–80 cm

Purple
80–90 cm

Orange
90–100 cm

Yellow
100–110 cm

Divot
110–160 cm

Red Clay
160–210 cm

Fig. 7.1. Construction of the layer-cake experiment. The colors of sand shown at the left were placed in layers in a plywood-lined pit. A harvester ant colony was then induced to dig a nest in this colored layer cake. From Rink et al. (2013).

ants could not roam about to forage, we fed them regularly with chopped mealworms, seeds, sugar water, and Pecan Sandies cookies. They seemed resigned
to their fates and went about vigorously excavating a nest in the layer cake. It
was a beautiful sight to behold as the ants piled brightly colored sand pellets
around the nest opening (fig. 7.2, *top*).

FIG. 7.2. *Top*, a harvester ant colony was released into a screen-bottomed cage (*left*) and excavated a nest in the layer cake below through a hole in the middle of the screen (*right*). The inset
shows a closer view of the pellets, many of which contain multiple colors of sand. The colors
and their amounts in the excavated sand revealed the depth and quantity of excavation activity.
The number of grains of each color told the story. *Bottom*, close-ups of sand collected from the
mound at *upper right*. Counting the grains in weighed samples of excavated sand revealed where
and how fast the ants were digging. Author's photos, from Tschinkel et al. (2015b).

The analysis of the pellets and the daily accumulation of colored sand told the story of the unseen creation of the underground nest in broad terms. Just as a house is built additively brick by brick, so is an ant nest created subtractively pellet by pellet. The scale of ants and the scale of soil granularity are not that far apart, so ants cannot use inertial forces to shovel sand into buckets for transport. Their interaction with sand is much more close-up and personal than ours, and we are hardly aware of the forces they must overcome. These include adhesion, surface tension, and the cohesion associated with thin films of water and with tiny points of cementation by minerals. At the scale of ants, water is a powerful glue. Each pellet is wrested individually from the "mine face" of a growing chamber or shaft by cutting, scraping, slicing, prying, or packing (fig. 4.7). The same forces that resist the formation of pellets also make them stable enough to be carried between the mandibles and dumped intact on the nest disk, where, with great care, they can be picked up with very light forceps and dropped into a well in a plastic plate with multiple wells. Jostled by wind and ants, the pellets fall apart as they dry to form the daily accumulation of sand. Every day, we collected 50 or so intact sand pellets, and then all the sand accumulated since the previous sample. Pellets or samples of a few milligrams were dropped into the wells in the well plates, and pictures of these wells taken with a digital microscope were later analyzed by counting grains (fig. 7.2, *bottom*).

The first appearance of a color on the surface told us the maximum nest depth on that day. From the bulk density of sand (1.5 g/ml), the mean number of grains per milliliter (13,000 ± 3,000), the mean percentage of each layer that was colored sand (25%; recall that we diluted the colored sand with native sand to save money), and the proportion of sand grains of each color (fig. 7.3), we calculated the relative rate of excavation in each colored layer, as well as the daily volume of chambers and their accumulated volume over time. Through it all, we and several undergraduate students counted grains of sand in hundreds of images of colored sand and entered these numbers in computer files. The data revealed a slow-motion movie in full Technicolor about what was happening unseen underground (fig. 7.3).

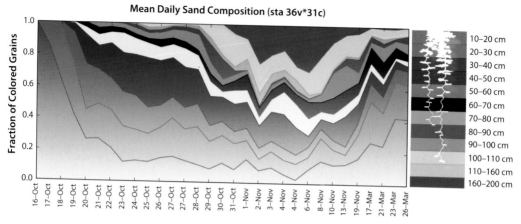

Fig. 7.3. *Left*, the fraction of each color represented in the daily sand samples revealed when the ants reached each layer, and the relative volume of chambers they dug in each. The volume of sand from the upper layers is much greater and appears earlier than that from deeper layers. *Right*, the sequence of colored layers, along with their depths. A schematic nest is superimposed on the column of layers for clarity. Modified from Tschinkel et al. (2015).

After seven months, we carefully excavated and mapped the layer-cake colonies, revealing the nest architecture level by level. By checking for colored sand residing in a host layer of a different color, we determined whether the ants deposited sand belowground without exposure to light, thus distorting the dates determined by OSL but also adding to our knowledge of the dynamics of nest formation. Moreover, by analyzing the color of such sand deposits, we could estimate the layer of origin, the distance moved up or down, and the amount moved and deposited.

The results were unequivocal—the ants deposited a lot of sand from deeper to shallower layers, and some even downward. The floors and walls of many chambers were lined with rainbows of colored sand, as were the walls of many shafts, some of which were completely backfilled (fig. 7.4). In the upper 70 cm of the nest, 90% to 98% of alien-colored sand came from below. Deposition was especially intense in the top 30 cm of the nest where many chambers were backfilled. About half of the 209 g of colored sand in the 10 cm thick top native layer was in the form of backfill. All in all, backfilling accounted for a

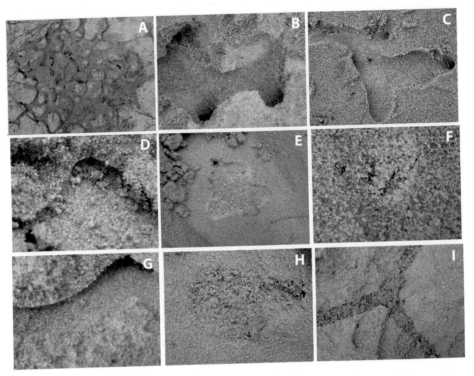

Fig. 7.4. Examples of the belowground deposition of sand from one colored source layer into another host layer of a different color. Deposits were smooth linings of chamber floors and walls, backfilled shafts or chambers, and loose pellets. Author's photos, from Tschinkel et al. (2015b).

greater share of deposition than chamber lining. Summed over all the nonhost sand deposited belowground, about 2.5% of the 13 kg of excavated sand was deposited belowground, about 75% of it in the upper 30 cm of the nest. If this were mixed homogeneously in the top 30 cm of nest soil, 4 grains out of every 1,000 would be alien to this layer. The contribution of such alien-colored sand to these deposits was directly proportional to the volume of chambers in these layers, and inversely proportional to the vertical separation of source and deposition. Interestingly, below 70 cm, up to 29% of deposits came from above (but these deposits were small).

For scientists who use OSL to date sediments, our results confirm that burrowing animals scramble the age-since-burial signal, and that this scrambling must be included in the interpretation of the data. As far as I know, there is no simple solution. One approach has been to date single sand grains and accept the youngest ages as the correct ones. This seems reasonable when the bulk of scrambling is upward. Like many areas in science, the OSL method is still a work in progress.

SAND TRANSPORT UNDERGROUND

The analysis of individual sand pellets revealed an important fact about how ants formed excavated pellets and transported them to the surface. If a pellet were formed and transported intact directly to the surface, it would contain only the color of the layer in which it was formed. But a large fraction of pellets contained multiple colors, meaning that they had been deposited in a host layer of a different color and then re-formed into a new pellet that incorporated sand from the temporary host. The number of colors in pellets ranged from 0 to 10 and averaged 4.4. Pellets contained an average of 162 grains (range 50 to 400), of which 60 were colored.

The degree of color mixing also tells us something about events occurring on the pellets' upward journeys. A pellet containing only one or two grains of a different color could easily be the result of simple contamination by an occasional ant fumble during transport, rather than of deposition and re-formation. If a second color forms a significant fraction of a pellet, it becomes much more likely that the pellet was reconstituted in a layer of a different color, and this becomes more and more likely as additional colors become larger components of the pellet. These trends can be described by the Gini coefficient, a statistic invented by the Italian economist Corrado Gini to describe the evenness of income distribution in an economy. Applied to the colored grains in our pellets, a Gini coefficient of 0 means that all the colored grains are of a single color, while at the other extreme, a coefficient of 1.0 is achieved when multiple colors are all represented by the same number of grains—that is, complete evenness.

As pellets contained more colors, their Gini coefficients increased—that is, the colors were more evenly represented. The most common coefficients were between 0.1 and 0.2 (fairly uneven), but more than half of the pellets had higher coefficients, with a few as high as 0.6 to 0.8 (quite even). Subsurface deposition of nonhost colors together with pellet analysis suggests that a substantial fraction of sand pellets are not carried directly to the surface but are deposited and formed into pellets en route multiple times, resulting in the observed frequency and evenness of multicolored pellets. In other words, sand transport takes place in a sequence of discrete steps, probably by multiple workers.

During our excavations of the layer-cake nests, we occasionally found intact pellets in chambers (fig. 7.4), but most deposited pellets were smoothed into the floors and walls of the nest. Pellets formed from these locations would obviously be increasingly mixed, but we cannot tell with assurance where most mixing takes place. What is clear is that the underground nests of harvester ants, once constructed, are not static but are being continuously remodeled. This dynamic nature is particularly obvious in the uppermost chambers, where backfilling and reexcavation seem to play a very large role. This continuous remodeling may or may not have direct adaptive value in producing a "perfect" nest (which probably doesn't exist anyway). It may be the consequence of workers responding directly to local cues and elapsed time, so that as the nest takes shape, the frequency of the relevant behaviors decreases but never completely stops. Just a little nudge here and there, or moving that pellet by 4 mm, or deciding that this is far enough, or not far enough.

MIXING ON A SHORTER TIMESCALE

But how dynamic is this process, really, in the short term? The deposits we saw in the layer-cake nests were formed over seven months. Could we estimate these rates on a shorter timescale? We could if we could induce a colony to suddenly dig into a layer of colored sand and we knew when they did so. We chose a more modest procedure by digging a pit half a meter deep next to an unsuspecting nest and then digging laterally until we just nicked a nest chamber. We then injected a few spoonfuls of pink fluorescent sand into this chamber, creating a mess that we knew the ants could not resist cleaning up. After one to three days, pink sand appeared on the nest disk, and chamber-by-chamber excavation of the seven nests showed that 27 of the 42 total chambers had pink sand, ranging from scattered grains to intact pellets. One of the uppermost chambers was backfilled with pink sand. Most chambers that lacked pink sand were at intermediate depths.

If we give the process some thought, it seems there ought to be a lot of incidental contamination resulting from workers carrying friable, fragile pellets upward through winding shafts and two-way traffic. I tested this in small, vertical glass "sand"-wiches in which the bottom 2 cm was fluorescent pink sand and the rest was native sand. Once the ants dug down into the pink sand, many of the pellets they carried upward were pink. Even the accidental loss of single grains from these pellets was detectable under ultraviolet light (fig. 7.5). In addition, ants often deposited pellets on their way up rather than outside. "Contamination" of the native sand was thus both apparently "deliberate" and accidental.

This experiment shows that deposition of sand transported upward from deeper in the nest is an immediate and normal part of the construction and remodeling of the nest, and although its rate may vary, it is a continuous and dynamic process. Had we excavated the layer-cake nests at different times, we would have found the nonhost deposits, while still present, to have been different in location and composition. It follows that when I make a cast of a nest, it is to some degree a snapshot in time.

COLLECTIVE SOIL MOVEMENT BY HARVESTER ANTS AT ANT HEAVEN

By now it is clear that each harvester ant colony moves a lot of soil around during nest construction. But how much soil do the 430 colonies on the 23 hectares of Ant Heaven collectively move on the scale of ecological or even geological time? We actually have enough information to calculate this movement. From the many excavations we did in 1985, from the relocation study, and from

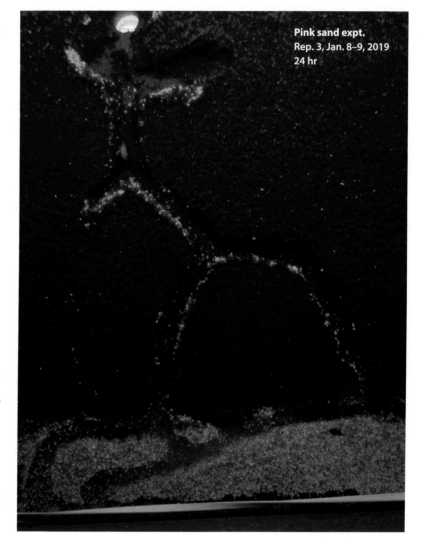

Pink sand expt.
Rep. 3, Jan. 8–9, 2019
24 hr

Fig. 7.5. Sand pellets are friable, so grains are often lost during upward transport of the fluorescent sand at the bottom. In addition, workers often deposit pellets short of the surface. Such "contamination" is constant during nest excavation. Author's photo.

the layer cake, we know how a nest's volume, depth, and vertical chamber-size distribution can be predicted from the size of its charcoal-covered nest disk. We also know how much soil is deposited belowground. From the repeated survey maps of Ant Heaven, we know the density and spacing of colonies, along with their nest disk areas, how far and how often they move, and finally, their life span (see chapter 4).

The measurement of disk areas in the surveys connects all these data because from them we can—without digging colonies up—estimate what the ants have done underground, including maximum depth, nest volume, and vertical distribution, and we can make these estimates over time. The estimated nest volumes at Ant Heaven ranged from less than 0.5 liter (about 0.75 kg of sand) to 12 liters (18 kg), with an average of about 3 liters (4.5 kg). Most of this excavated sand is dumped on the nest disk, where wind, rain, and animals disperse it quickly, so that after a few years it has been spread evenly to form a uniform layer.

So let's plug our data into a computer simulation of what the ants do over time, starting with an initial virtual "colony" at a random point in the average space occupied by each colony (670 m²). The colony "moves" once a year to another random point 3.9 m away (± 3.15 m). After 20 years (±4 years), the colony "dies" and its territory is taken over by a new, small colony "founded" at a random point within this space. The new colony grows to its "mature size" of 3.3 liters (±0.6 liters) in 6 years (±2 years) and moves annually as did the first colony, excavating nests of a size estimated from real volumes and disk areas. The vertical distribution of chamber volume reveals the amount of sand excavated from each 50 cm depth increment.

By the end of a millennium, each "colony" would have brought about 2,660 liters of soil to the surface and covered 21% of the area with disks. Most of this sand would have come from the shallowest depth increment (0–50 cm) because that is where most of the nest volume is located. In a millennium, the 16 colonies occupying each hectare would have deposited 64 metric tons of sand on the surface, with 1.2 tons (2%) coming from depths greater than 2 m. When spread evenly, this soil would form a layer 0.43 cm (±0.05 cm) thick.

The output of this simulation can be visualized as a set of images showing the location and size of colonies after a given number of moves (equivalent to years). The depth source of the disk soil is coded by color, with shallow soil being blue and transitioning through green and orange to red as an increasing amount of soil comes from greater depths (fig. 7.6).

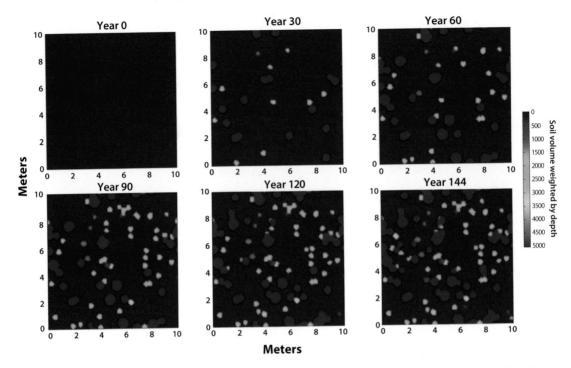

Fig. 7.6. An example of a simulation of biomantling by the nest disks of one colony and its descendants in 100 m² over 144 years. The mixture of soil from greater depths is coded by color, with blue indicating predominantly shallow depths and increasing red an admixture of deeper soil. From Tschinkel (2015b).

A BIGGER PLAYER IN BIOTURBATION

Pogonomyrmex badius is thus a significant agent of biomantling and bioturbation, but it is far from the only ant actor in this process. The Atlantic and Gulf Coastal Plains are home to a rich fauna of ground-nesting ants, some of which, such as the winter ant, *Prenolepis imparis*, make much deeper nests than *P. badius*, while others make shallower but much more abundant nests. A conspicuously abundant ant in the sandhill habitat is the northern tuberculate fungus gardener, *Trachymyrmex septentrionalis*, whose colonies rarely top 1,000 workers but whose nest density can be 1,000 per hectare or higher, or more than 60 times the density of *P. badius*. In spite of their small colony and nest size, these ants therefore bring 1.0 metric ton to the surface every year, burying this landscape under 6 cm of excavated soil every millennium. This rate is 15 times as high as that of the far larger colonies of *P. badius*, showing that abundance can more than make up for small colony size. On the other hand, *P. badius* digs far deeper nests, moving soil from depths of a meter or more. The contributions of other ants in this habitat are currently unknown, but each species contributes in proportion to its nest density, volume, depth, relocation frequency, and life span.

So far, I have focused mostly on biomantling, but as we know from the layer-cake experiments with *P. badius*, about 2.5% of the excavated soil is deposited belowground, mostly in the top 30 cm of the nest (see above). This amounts to 1.6 metric tons per millennium, not very impressive, really, but a consequence of low nest density. But the biomantling rate of *T. septentrionalis* is 15 times that of *P. badius*. If some proportion of this soil is deposited belowground, then *T. septentrionalis* might be a much bigger player in bioturbation than *P. badius*. Because of the success and fun of the layer-cake experiment with *P. badius*, we undertook a similar experiment with *T. septentrionalis*. My former student Jon Seal and I had already observed loose backfill in chambers, especially after late summer. Moreover, the nests contained only a few egg-shaped chambers to house their fungus gardens, and these were located at very discrete levels, ideal for the study of backfilling (fig. 7.7). Few nests were deeper than a meter, and digging them up was much less daunting than digging up a *P. badius* nest.

This time we used only five colors in 20 cm thick layers, with pits 1 m deep and 50 cm in diameter. Before we layered the soil, we lined the pit with a polyester fabric so the ants could not wander off the reservation and excavate outside the colored sand column (which they did in a pilot run). In mid-April, when colonies were just starting their fungus gardens and were still broodless, we dug up 10 *T. septentrionalis* colonies with queens and released them into screen-bottomed cages with a central hole through which they could dig a nest in the layer cake, which they all did within a day. Because the colonies could not forage, Nicholas Hanley or I added substrate for the fungus gardens two or three times a week and at the same time collected the soil deposited in the screen

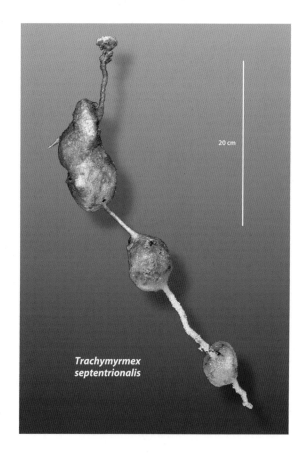

20 cm

Trachymyrmex
septentrionalis

FIG. 7.7. An aluminum cast of the nest of the tuberculate fungus gardener, *Trachymyrmex septentrionalis*,
showing the egg-shaped chambers that house its fungus gardens. Author's photo, from Tschinkel (2015a).

cage. The relative representation of the colored sand grains was determined
from photographs of these samples, as we had done for the *P. badius* layer cake.

Over the six months of the experiment, each colony brought an average of
760 g (range 460–1000 g) of sand to the surface, 82% of which came from
colored layers. By late May, this sand revealed that colonies had excavated 200
to 650 ml of chambers, mostly between 30 and 70 cm belowground. Thereaf-
ter, there was little excavation until late summer, when the ants dug deeper
chambers and deposited much of the excavated soil in shallower chambers, as
we will see below.

In late November 2015, we carefully excavated and censused the layer cakes
(fig. 7.8), collecting all chamber contents, including ants, fungus garden, and
backfilled sand. Analysis of the backfilled sand colors revealed that colonies

Fig. 7.8. Excavating a *Trachymyrmex septentrionalis* layer-cake nest. The layers of colored sand in which the ants dug their nest are visible in the front of the pit. Nests waiting to be excavated can be seen in the background. Author's photo, from Tschinkel and Seal (2016).

moved an average of 150 g (range 46–330 g) of sand from one colored layer to another. Combining this 150 g of underground deposit with the 760 g they deposited on the surface, we see that about 17% of the total sand was moved and deposited underground. The direction of movement was not random but was mostly upward, with most backfill less than 70 cm belowground. All layers received more deposits from below than above, mostly from one to two layers

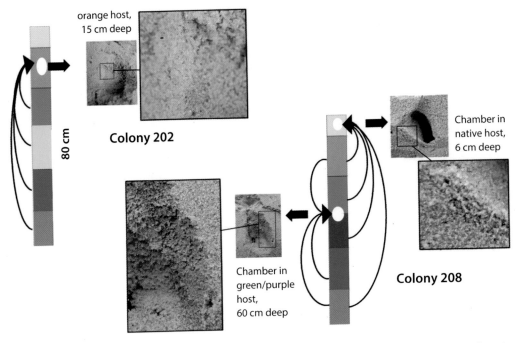

orange host,
15 cm deep

Colony 202

80 cm

Chamber in
native host,
6 cm deep

Chamber in
green/purple
host,
60 cm deep

Colony 208

FIG. 7.9. Three examples of belowground deposition of colored sand from one source layer into a chamber in a host layer of a different color. The sequence of colored sand layers is shown in the bars. Arrows indicate the source and deposition of colored sand from the source layer to the chamber. Images show the deposition chamber along with a close-up of the deposited sand. Author's photos.

lower. Many of the three to six chambers in each nest (n = 9 nests) were partially or completely backfilled, usually with sand from more than one source layer (fig. 7.9). There were backfilled shafts, lined chamber floors or walls, layers of single colors, and completely backfilled chambers. Deposition in layers suggested that the ants dug in each source layer for a time and then switched to another layer, a conclusion that was also supported by a chamber that was filled from the bottom up with different colors (fig. 7.9).

Although it is clear that *T. septentrionalis* deposits a lot of soil both on the surface and belowground, the fate of these deposits is probably very different. Wind, rain, and animals spread the surface deposits into a more or less uniform layer over the years, but the belowground deposits probably do not mix and spread rapidly. Construction of these little nests is localized in a cylinder of soil 50 cm or less in diameter, so that even if the colonies move into a new nest annually, they will bioturbate only the soil below about 200 m^2, or 2% of a hectare, in a millennium. Whereas soil may seem to be a pretty homogeneous medium to a person with a spade, to small creatures and plants, it may be a mosaic of highly variable patches.

If we compare bioturbation by the endearing little David *T. septentrionalis* with that of the Goliath *P. badius*, we see once again that abundance can more than make up for small size. *P. badius* deposits about 2.5% of excavated soil belowground, and *T. septentrionalis* deposits 17%. On a per-nest basis, this amounts to about 110 g for an average-sized *P. badius*, and 153 g for *T. septentrionalis*. But with 1,000 or more nests of *T. septentrionalis* per hectare, this little fungus gardener collectively stirs 153 metric tons per millennium per hectare, or 96 times as much as the seed harvester. It also biomantles the surface with almost 15 times as much excavated soil, but in contrast to *P. badius*, very little of this soil comes from deeper than a meter.

A FINAL NOTE

The nest architecture of a colony of *P. badius*, with its large size and aesthetic appeal, is created over and over, year after year, in brief moments of the colony's annual cycle. The nest is not the result of the slow enlargement of a tiny founding nest. Whereas *P. badius* moves a lot of soil, the example of *T. septentrionalis* shows we should not judge effect from size only. *P. badius* and *T. septentrionalis* suggest how the 30 to 50 common ground-nesting ants that make up this ant community churn the soil to undetermined depths and degrees, with ecological effects that are only dimly visible. Although there are a few studies of the biomantling of selected ant species in a few habitats around the world, the collective effects above- and belowground are largely undetermined. But as long as ants dig nests in the world's soils, and as long as they move into new ones now and then, they will continue to be important players in rejuvenating the world's soils, for a soil left in place without intervention for a very, very long time slowly dies as its mineral wealth is weathered, leached, and transported to sea in rivers and streams until it has little left to offer plants. Nutrients lofted from the sea or dust deposited in rain can slow this process but cannot stop it. Dissolved essential nutrients that are not captured in time by plants continue their downward journey until they are out of reach of plant roots, gradually seeping into streams that eventually carry them to the sea. In the course of building their nests, thousands of ant species churn the soil and slow these losses by bringing soil and its nutrients back up into zones where plants have access to them. The world's ants thus play an essential ecological role in maintaining soil fertility in most of the warm regions of the world.

CHAPTER 8

Division of Labor and the Superorganism

THE EVOLUTION OF BIOLOGICAL COMPLEXITY

There is a "wholeness" about an ant colony. It is an entity with sharp boundaries and "personal characteristics," and with a clear internal structure within its three-dimensional nest. Its many organism-like traits, along with recent advances in evolutionary theory, have given new life to the metaphor of a "superorganism."

To understand the origin and insightfulness of this metaphor, we have to go back to a time when life on Earth began as self-replicating molecules in the primordial soup of the warm newborn oceans, an era of which we know little. To say that life's complexity has increased in the last three billion years is a vast understatement. While some of this increase occurred through gradual natural selection, all the major leaps in complexity ("quantum leaps") arose through the symbiotic combination of the most complex life forms available at the time. Between 3 and 4 billion years ago, diverse molecules combined cooperatively into units we now call bacteria, a sort of molecular soup enclosed in a cell wall and membrane in which the previously independent molecules became the subunits of this new, much more complex entity. Next, between 1.5 and 2 billion years ago, several different kinds of bacteria combined symbiotically to become the subunits of a radically more complex life form, the eukaryotic (nucleated) cell. The formerly independent bacteria became membrane-bound cellular organelles, some even retaining their own DNA. Between 1 and 1.5 billion years ago, eukaryotic cells combined symbiotically to become multicellular plants and animals whose descendants include ourselves.

As in the earlier transitions, previously independent single cells became the subunits of a radically more complex and integrated entity.

These transitions in complexity have several things in common. In each transition, each of the previously independent entities became the subunits of the new entity, with each subunit specializing in a different life function, but all sharing the goal of reproducing the entire entity. In each case, subunit reproduction mostly built the new entity, and the crucial reproduction moved up a level to the entity as a whole. Molecules, organelles, and cells reproduce (or are replicated) within the confines of (respectively) bacteria, eukaryotic cells, and multicellular organisms, tying their future completely to the entity of which they are part. With each jump in complexity, mechanisms for integration and coordination of the previously independent parts had to evolve, becoming increasingly complex and higher level as the life form became more complex.

The appearance of multicellular organisms such as animals brought with it a fundamental change in reproduction. In the unicellular bacteria and eukaryotes, all the subcellular components and organelles were simply replicated and divided between the daughter cells during cell division as a sort of captive population. But in an animal consisting of thousands to trillions of cells, this was no longer possible. Interestingly, the reproduction of animals takes place through reversion to the single-celled ancestral condition—that is, through the production of single-celled gametes (eggs and sperm). These combine during fertilization and then proceed to produce the new organism through repeated cell division and differentiation. As these cells multiply, they change structure and function to become many cell and organ types through complex cascades of gene activation and inhibition, cell migration, and cell death. The final outcome is an animal in which the diverse life functions are carried out by differentiated cells, tissues, and organs. Even then, the gametes are not produced by a bunch of randomly chosen cells, like a lottery, but by a line of cells (the germ line) that are set aside very early in development and that do not participate in forming the body of the developing animal. Only the germ line cells can pass genes on to the next generation by producing eggs and sperm. All the rest of the cells in the body (somatic cells) have no reproductive future save through ensuring the reproductive success of the germ line cells. But of course, all are genetically very similar, having all been produced by mitotic cell division, so all have a stake in the success of the animal of which they are part.

With each leap in complexity, the primary target of natural selection moved up a level to the new combination—from molecules to bacteria, from bacteria to eukaryotic cells, from eukaryotic cells to multicellular organisms. Through the specialization and modification of the subunits, the appearance of each new level of organization was followed by an explosive increase in diversity based on the new body plan. Because the modern representatives of each of these levels are so tightly integrated and honed, their origin as independent entities is no longer obvious.

THE SUPERORGANISM ARISES

The final step in this multiplication of complexity is less obvious, in part because it is less complete, and in part because features that made it similar to previous leaps took longer to recognize. About 100 to 200 million years ago, previously solitary insects evolved sociality, adding a still higher and radically more complex level of biological organization to the living world, an organism of organisms, so to speak—the *superorganism*. There are many parallels between a superorganism such as an ant colony (or other social insect colony) and an organism—just as the cells in an organism have no genetic future save through helping the germ line cells produce offspring, so the workers in an ant colony have no genetic future save through helping the queen and her germ line produce more offspring colonies. Animals reproduce by reverting to a single-celled ancestral condition (eggs and sperm). In a parallel manner, most ants also revert to their ancestral solitary condition (sexual females and males). Just as the offspring animal is then produced by the multiplication of somatic cells, so the daughter colony arises through the addition of nonreproductive workers. Just as genes pass to the next generation only through an animal's germ line cells, so do they pass only through the sexual ants and their germ lines. Just as the essential functions of life are carried out by specialized organs and tissues in an animal, so these functions are carried out by specialized groups of ants in an ant colony. Both animals and superorganisms have complex mechanisms for coordination that ensure harmonious operation of the whole, and both undergo differentiation into specialized parts during growth. At all levels, this differentiation is guided by activation and inhibition (a newly developed item activates or inhibits the development of another), feedbacks (positive or negative), and movement. The final superorganism is no more like its initial state than an adult animal is like a zygote (perhaps an exaggeration, but you get the point).

One of the superorganism's characteristics that slowed the recognition of parallels is that unlike the parts of the organism, those of the superorganism (the ants, for example) are not glued together into one piece but are free to move about in the environment. Even within the nest, they move about within and between their chambers, licking larvae, carrying food, or just standing as though in a stupor. This apparent independence somewhat obscures the fact that a nonreproductive worker is but a small part of a much larger self-reproducing entity and has no reproductive future on her own. I should caution that workers should not be regarded as the literal cells of the superorganism, nor groups of workers as its organs. Such characterizations brought the first iterations of the superorganism in the early twentieth century into disfavor.

DIVISION OF LABOR IN ANTS

The specialization of the parts of organisms for particular functions is at the heart of biological organization at all levels. This subdivision and specialization of functions in superorganisms largely takes the form of what is usually called division of labor. Similar subdivision and specialization of functions is a central part of most forms of organization, be they corporations, armies, religious organizations, multicellular or single-celled organisms, or social insect colonies. Closer consideration of human organizations that make things (factories) is instructive because insect colonies also make things. In a factory, the many tasks needed to construct the final widget are assigned to different workers or groups of workers, who then become very proficient in their part of the process, contributing to the efficiency of the whole factory. If every individual worker carried out the whole sequence of widget-producing steps, widget production would be much less efficient. Henry Ford figured that out a long time ago.

Social insect colonies are factories that convert the raw material of food into workers as an intermediate step for producing sexual ants, which in turn will produce daughter colonies. To this end, social insect colonies, like human factories, are universally organized so that different individuals or groups of individuals specialize in different parts of the whole enterprise. To call this "division of labor" is inadequate and somewhat misleading because it seems to refer only to differences in behavior—that is, to work. In reality, even the most basic division of "labor" (reproductive and nonreproductive; fig. 8.1), the one that actually defines insect sociality, is actually mostly physiological—one individual (or a few) in an ant colony (the queen or queens) has all the necessary apparatus and physiology for producing and laying fertilized eggs (ovaries, fat bodies, spermatheca), mating (spermatheca), dispersing (wings and wing muscles), and colony founding (fat and protein storage). In contrast, workers are stripped-down ants that lack most of these features and are more or less sterile, carrying out most of the remaining colony functions. Clearly, the laying of eggs by the queen is a behavior, but it is only the final step in a long and complex series of developmental and physiological steps. As part of this function, the queen is also responsible for colony regulation through the secretion of pheromones and other materials that influence her workers. Again, this is not physical work.

As myrmecologists have explored the reproductive biology of an ever greater number of ant species, they have found variants and exceptions to even this simple and basic functional division. These range from egg laying and reproduction without fertilization by workers (parthenogenesis) to multiple queens, queenlessness, and workerless social parasitism. These are outside the focus of this book, but in any case, they describe only a small proportion of the life histories of ants, most of which represent the "ordinary" queen-worker dichotomy as described above.

FIG. 8.1. Worker fire ants surrounding their queen. Photo
from Tschinkel (2006).

The "work of workers" has received a great deal of attention in social insect
research. Much of this has focused on how workers do their various jobs and
how labor is self-organized, as well as the stimuli that activate their various
behaviors and the efficiency with which they do what they do, be it gathering,
processing, and distributing food; converting it and storing it as fat; excavating and maintaining the nest; defending territory and resources; or caring for

the brood and queen. Even so, not everything ant workers do is behavioral, for depending on the species, they may specialize in storing fat or other food, carry out digestion whose products are shared with nest mates, be responsible for other metabolic processes, or simply be a quiescent group in reserve for later functions. Such functions are not visible as behaviors but are important parts of colony function, nevertheless.

Research on a large number of ant species has shown that there is a universal division of labor and function by worker age—young workers function mostly in brood care, change to more diverse nest duties as they age, and finally become foragers outside the nest during the last phase of their lives. As foragers, they usually do not live very long, at least in nature, for foraging is dangerous because it exposes workers to predators, desiccation, overheating, enemies, and getting lost. These changes in what workers do are associated with changes in their physiology, hormones, and nervous system. Worker ovaries, if they have them, generally regress as they age, as do the levels of juvenile hormone and fat stores. Through age and experience, their behavioral repertoire increases, as do the sizes and connectedness of relevant parts of their brains. It is perhaps not surprising that the foragers, which face myriad challenges and dangers out in the world, need a lot more brain power than do brood tenders deep in the protected confines of a moist nest chamber. Simply put, foraging workers need to be smarter, and they become so as they age. A currently debated question is how flexible (or rigid) these age-associated behavioral shifts are. The extreme points of view are that the transitions are flexible and respond to colony demand, and alternatively that they are demographic—that is, determined mostly by worker age and thus relatively unresponsive to demand. I will have more to say on this later.

There is one additional basis for division of labor: worker size (fig. 8.1). About 15% of ant species have evolved workers that vary greatly in size, either continuously from small to large, or with two (occasionally three) distinct and nonoverlapping sizes (minor and major workers). Large and small workers are specialized on different task sets. The smaller workers usually make up a larger proportion of the worker population and exhibit a wider range of behaviors.

In contrast, large workers are usually more specialized on tasks that occur more rarely. In some species, for example in *Pheidole* spp. and *Pogonomyrmex badius*, the large workers have disproportionately large heads and mandibles, suggesting functions of crushing, chewing, or defensive biting. Many authors casually call such large workers "soldiers," but in most cases their military specialization has not been firmly established. On the other hand, in many species, such major workers participate little or not at all in brood care. For example, when the largest workers in a fire ant colony are put in charge of rearing larvae, the larvae die from neglect. On the other hand, these large workers are disproportionately active in foraging.

All workers go through these "labor phases" as they age, but the age at which they make these transitions varies greatly among species and across seasons, as does the proportion of life spent in each phase. Natural selection adjusts these transitions to suit ambient conditions and life habits of colonies. For example, when the natural history requires more foragers and fewer brood workers, the transition to forager occurs at younger ages.

A final division of labor applies to only a small proportion of the workers that carry out the same specialized and relatively rare task for a long time. These are sometimes referred to as "career workers." They include trash workers in leafcutter ants, "tanker" ants in honeydew-harvesting thatch ants, undertaker ants in several species, and "shuttle" fire ant workers that collect fertility-stimulating material from metamorphic larvae and feed it to the queen.

A relatively little-studied area is whether and how the division of labor changes as colonies grow. When division of labor is based partly on worker size, the colony's labor force shifts from predominantly brood care and general nest duties to foraging as a result of large colonies producing a larger proportion of large workers. Both of the primary kinds of division of labor develop as the colony grows, with the most rapid shifts when the colony is small. Even for a given size of worker, task specialization increases with colony size. Thus, we can expect the pattern of division of labor to depend not only on the ant species but also on the colony size and perhaps age.

Because division of labor is a hallmark of the superorganism, we can expect that creation of the nest also involves division of labor, in particular with respect to nest site choice, initiation of excavation, excavation itself, transport of soil, and even the depth at which workers excavate and transport. Through this spatial and behavioral division of labor, the superorganism creates the space in which it lives, the analogue of its "body" (to the extent that superorganisms have bodies). Importantly, the particular architecture of a species is due largely to differences in the details of its division of labor.

SUPERORGANISM AS THE "FOREST" RATHER THAN THE "TREES"

Like all biological studies, the studies of animals as organisms began with descriptions. By the early 1990s, I was convinced that we lacked basic descriptive knowledge of even the most frequently studied ants. Rarely could myrmecologists who were testing complex and sophisticated hypotheses report basic attributes like colony size, colony or worker life span, seasonal patterns, colony growth and development, and so on. Think of such descriptions as the reenactment of *The Anatomy Lesson of Dr. Tulp*, which Rembrandt painted in 1632, but on social insect colonies. In the painting, Dr. Tulp has partially dissected the arm of a cadaver and is pointing out essential anatomical features while his students look on attentively. The idea is that each part of an animal (including Dr. Tulp's dissected human) has a characteristic size and shape that arises during development, is shaped by evolution, and often changes with life stage and even season. All these are adaptations to deal with the demands of being alive, demands that both unitary animals and superorganisms must meet. All further insight into how animals work begins with a description of their anatomy. We therefore need to ask questions like the following: What are the counterparts of such functional anatomy in superorganisms? What is their "size and shape"? How do they arise during the superorganism's development? How are they shaped by evolution? How do they change with life stage and season? By analogy, the question is not how the liver of the mouse organizes itself to do its job, but how much of the mouse's total resources are allocated to the liver, and what factors affect this allocation. Applied to the superorganism, the question is not how foragers or brood care workers organize themselves to do their jobs, but how much of the colony's resources are allocated to foragers or brood care workers, and what factors affect this allocation. The focus is thus on a higher plane than the individual and her competence; it is on what proportion of the colony's labor and resources is devoted to her *kind* of job; how this allocation changes with colony size, niche, and season; and how it contributes to colony fitness.

RESOURCE ALLOCATION IN THE SUPERORGANISM

The reason this is a useful way of looking at both organisms and superorganisms stems from economic thinking—resources are limiting for all organisms and superorganisms, so resources allocated to one organ or labor group are unavailable for allocation to another, and, much as with investing money, there will be an optimal allocation pattern that maximizes reproductive (or monetary) success. Therefore, just as natural selection adjusts, say, the size of the liver to an optimum, so should it adjust how much worker labor, time, and energy are allocated to each superorganism function to achieve an optimum. This principle applies at every level—molecule, organelle, cell, tissue, and organ. In biology this optimum is not necessarily constant but can depend on life stage, environment, season, and other factors. In large measure, the diversity of organisms and superorganisms is the result of different patterns of allocation.

With this background, we can now see that a parallel description of a superorganism requires that we identify the different groups within the colony that carry out its major life functions. Unlike in a unitary animal, there is no bloody blob of a liver that we can weigh—no baggy lungs, no throbbing heart and hollow arteries. The differentiated parts of an ant colony perform their diverse functions without being attached to each other. Much of this functional differentiation is behavioral division of labor, but it involves much more than behavior, as I noted above. Having identified the major labor groups within the superorganism, we need to determine what proportion of its resources is invested in each, for patterns of this investment are the parallel of a unitary animal's investment pattern in its organs. These patterns of allocation are expected to vary with life history stage, ecological niche, season, and other such factors and can be viewed in economic terms.

The second major task is beyond the scope of this book, but it is to recognize that the superorganism, like the organisms that make it up, has a life cycle—zygote/founding, growth, maturation, reproduction, and death. This means that each species of superorganism has a characteristic mature size, life span, and reproductive output. Moreover, it develops from founding to maturity by means of the rules and interactions of "sociogenesis," much like the unitary animal develops via those of ontogenesis. The characteristics of both change with the stage of development. In the ant colony, much of the change will involve shifts in the relative size of the labor sectors, but also major shifts in worker morphology and physiology. Relatively little is known about this subject, and a serious treatment will have to wait for a future book.

ORGANIZATION OF LABOR IN SPACE

As we saw in chapter 3, the ant colony exists in a three-dimensional space in which the ants are not randomly distributed. Optimal efficiency also requires not only optimal allocation of labor, but also an organization of tasks in space so that work flows from task to task in an order demanded by the assembly of the product. Henry Ford would never have put the workers who installed car seats next to those who tightened the head bolts on engine blocks. Task specialization and spatial organization obviously go hand in hand in factory assembly lines, and we can expect that it does so in insect colonies as well. That this spatial component is important has largely escaped social insect researchers, but this is exactly where nest architecture intersects with division of labor. It is strange that this intersection has been so rarely recognized, but it is perhaps because almost all studies of division of labor in ants have taken place in simple laboratory nests that bear no resemblance to natural nests, so the question never even comes up. Perhaps for reasons of convenience, ant researchers have rarely questioned whether this might cause them to miss something important in division of labor. In retrospect, anyone who has kept pet ant colonies knows that the ants very quickly organize whatever space they are given so that pupae and other nonfeeding brood are piled in one area, feeding larvae are spaced out nearby, the queen is surrounded by her attendants in a third area, and the workers that receive food from incoming foragers huddle near the nest entrance.

Marking and tracking individual workers reveals that there is limited exchange of workers among these self-organized zones, showing that even in the artifice of a laboratory nest, organization in space is a consistent component of division of labor. For example, the tiny colonies of the ant *Leptothorax unifasciatus* nest in the thin spaces between stones. When set up between plates of glass in the laboratory, individually marked workers moved only within limited concentric zones, each of which was associated with a different kind of labor. Workers in the central zone were more likely to take care of brood, and those on the periphery were more likely to forage. Importantly, when the ants moved or were forced to renest, they re-created the same precise arrangement in the new nest. To these ants, space and work are closely associated.

THE FLORIDA HARVESTER ANT AS A MODEL SUPERORGANISM

If ant labor is spatially organized even in the cramped conditions of a laboratory nest, we should expect that it will be much more segregated within a natural nest of the Florida harvester ant that stretches for two or three vertical meters. Early evidence of this came from a number of field studies of natural nests by Bill MacKay and Sanford Porter, who found that in several species of *Pogonomyrmex*, workers collected from deep in the nest were mostly younger, as evidenced by their paler color (see chapter 3). In Japan, Masaki Kondoh dug up very deep colonies of *Formica japonica* and found that the young, fat workers were limited mostly to the deeper parts of the nest. I found a very similar distribution of young, fat workers in the 4 m deep nests of the winter ant,

Prenolepis imparis. Because the association of young workers with brood care was well known, it seemed likely that young workers deep in the nests of natural colonies were also involved in brood care.

My description of colonies of the Florida harvester ant at the superorganism level began in 1989, when I set out on a loony project to excavate, capture, and census six complete colonies (two small, two medium, and two large) during each of four seasonal phases of one year. Colonies were often 2 to 3 m deep, and excavating them in the heat of summer (or even May or October) was enough to make me and my assistant Natalie Furman dizzy. As each chamber was exposed by lifting off the sand with a brick trowel, a piece of transparent acetate was placed over it and its outlines traced with a marking pen (see chapter 3). By the end, we had dug up 31 colonies.

The outcome of all this digging and collecting was a mountain of vials, bags, bottles, and dishes containing the dead and dried workers, larvae, and pupae (by chamber), as well as the seeds (also by chamber). We separated the workers by color (dark = old; pale = young) and counted them and the brood. Each ant in a sample of 100 workers was weighed individually, followed by removing the head and measuring its width as a correlate of body size. Seeds from each chamber were sieved into 10 sizes, each of whose average weight was determined. This resulted in a huge pile of data that I massaged on a mainframe computer for over a year to figure out the patterns of distribution of nest contents within the nest space and by season. As in chapter 3, analyzing the chamber tracings revealed much about nest architecture, including the distribution of chamber size and shape by depth, colony size, and season, as well as total chamber area, total volume, and maximum nest depth for each colony.

At the time I did this work, computers had recently made it possible to analyze these data chamber by chamber, but these patterns were often rather noisy, so I aggregated data in several ways to reveal average patterns—by tenths, thirds, or quarters of nest depth; by colony size category (small, medium, large); and by season (January, May, July, September–October). This large data set allows me to add a good deal of nuance and detail to what we already saw in our dig in chapter 3, that workers are strongly sorted by age within the nest.

The study also showed a wealth of other characters, but these are less relevant to the present discussion.

Callows (young workers) were present in all seasons except May, suggesting that colonies do not produce new workers in the winter—and indeed, brood were absent in January. The callows predominated greatly in the lower nest regions, presumably engaged in brood care, as in most ants (fig. 8.2). And indeed, the great majority of callows occurred mostly in the same nest regions (the bottom third) as larvae and pupae (as we saw in our dig in chapter 3). We can reasonably designate the lower nest regions as the colony's crèche, with young workers acting as nursemaids for their sister brood. While pupae are sometimes brought into the uppermost chambers to warm them and speed their development, this is always a small proportion of the total brood, most of which remain deep in the nest.

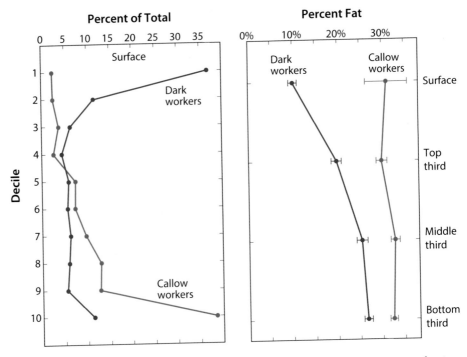

FIG. 8.2. Workers are fatter when young, losing fat as they age. Young workers occur predominantly in the lower regions of the nest.

For a given body size (estimated from head width), recognizably callow workers were heavier than darker ones, and extracting the fat from these workers with ether showed that most of the weight difference was fat. No matter at what level callows were captured, their "fatness" changed little (from 33% to 26%), meaning their fatness was associated mostly with their youth, not their location in the nest (fig. 8.2). As they gradually darkened and aged in the bottom third of the nest, their fatness remained similar to that of callows (30% and 33%, respectively), even if they sometimes strayed to higher levels in the nest. On the other hand, as these eventually mature dark workers moved to higher levels in the nest, they lost weight by using up their fat reserves, finally ending as foragers on the nest surface with only about 12%–13% fat, down from their initial reserves of 30% (fig. 8.2). The colony thus allocated foraging, the most dangerous task, to the most "used-up" workers and thus minimized the loss of colony resources upon the inevitable death of such workers. Ants, in other words, send their old ladies out to do the dangerous work.

One might reasonably think that this vertical distribution by age is just a meaningless epiphenomenon of ants being born in the lower nest regions and drifting upward, but this is not the case. I constructed an "ant hotel," a set of chambers on a 1 m central shaft to which the ants could be given or denied access by turning a tube within the main vertical shaft, like a stopcock. With the stopcock closed, each chamber in this meter-tall hotel was stocked with marked workers, half of them young and half old, along with a few larvae, and buried up to the top in the ground. After a day's acclimation, I opened the stopcock so the ants could assort themselves at will, after which I closed the stopcock, extracted the hotel, and counted the ants in chambers. The young workers had moved little, but most of the old workers had moved upward into higher chambers. Thus, not only is the vertical age social structure "deliberate," but the ants "know" where they are in vertical space. We touched on this point in chapters 5 and 6.

We thus have two age points in the division of labor—young workers carry out brood care deep in the nest, while the oldest workers leave the nest to forage. In between these bookends, it appears that on average, the ants move upward as they age, probably changing jobs as they do. What these job changes might be is discussed below, but first let us inspect the job of forager more closely, for this is the only labor we can observe directly without disturbing the nest. It is also the only labor group we can define operationally and unambiguously—any worker that travels some distance from the nest entrance, picks up a food item, and carries it back to the nest is a forager. This unambiguous identification of foragers allows us to capture as many as we have patience for or as the colony can field, making it possible to study many aspects of this distinct subpopulation.

ESTIMATING POPULATIONS WITH MARK-RECAPTURE

Because our goal is a quantitative description of a representative superorganism, the most obvious questions are (1) How many foragers are there, and what proportion of the colony forages? (2) How is the forager population related to colony size? (3) How does the forager population change with the seasons? and (4) How are the foragers distributed within the nest? Between 1996 and 2005, I had done a very successful study of this type on fire ants, and I now suggested to my student Christina Kwapich that she do a similar study on other ant species.

In order to understand how Christina answered these questions, we must digress briefly to describe the method for estimating the population size, the so-called mark-release-recapture method. Let's say that for some reason you need to know how many animals are in a population. Obviously, if you could see all the animals at once you could simply count them, but how likely is that? Your "animals of interest" are probably scattered over some considerable territory rather than bunched together where you can count them. OK, you could put a mark on each animal the first time you see it and keep track of the marks until you see no more unmarked animals. How practical is that? But it does suggest a variant that might have a future. How about if we capture as many animals as we can in one session, mark them, let them go again, and then a day or two later capture another big bunch and count the proportion that bears a mark? All we then have to do is assume that the initially marked and released group is the same proportion of the total population as the number marked is of the recaptured group. For example, say we captured and marked 100 animals and released them, and the next day we found that 10% of the recaptured sample was marked. This means that 100 is 10% of the total population, and there are 1,000 animals in all.

Simple, right? But of course, nothing is that simple. Well, actually it is, *if* (1) the population is bounded—that is, the animals roam within a definable area; (2) the marked animals mix randomly with the rest of the population; (3) there is no emigration or immigration; (4) there are no appreciable deaths or births over the interval between marking and recapture; (5) marked and unmarked animals are equally likely to be captured; (6) marking does not affect animal survival; (7) the mark is not lost at a significant rate; and (8) a substantial proportion of the population is marked. Some of these issues can be partly addressed with more complicated procedures, but most of them dog all studies of typical mobile animals.

This doesn't bode well, does it? After all, animals wander wherever they please rather than stay in bounded areas, they die, they migrate, new ones are born, and marks are not permanent. Back in the early 1970s, my friend John and I undertook a mark-recapture project on darkling ground beetles in the Arizona desert. Every night, we captured bunches of these large black beetles, numbered and marked them individually, and let them go again. Our recapture rates never got within even shouting distance of 5%. The beetles' behavior upon release should have tipped us off—each chose a random direction and headed straight for the horizon, never to return. This population had no bounds—their home was as large as the desert. Many studies of mobile animals have similarly low recapture rates, which means that the population estimates derived from them are very uncertain—say, based on the recapture sample, there is a 95% chance that the population is between 150 and 10,987. Well, that certainly pins it down! Intuitively, we can see that the higher the fraction of the population we mark, the more accurate our estimate will be. When we have marked 100%, what we have is no longer an estimate but a simple and accurate count.

But one group of animals is very well suited for mark-recapture studies: the ants. Or more specifically, the ant foragers that wander about on the surface. The reason is that the foragers of each colony form a closed, bounded population because they occupy a defined area and do not mix with neighboring forager populations—they hardly mix with their own colony members in the nest. This condition has been shown to apply to practically all ants that have been studied in enough detail—all have inside ants and outside ants (foragers). True, they are only a portion of the whole colony, but they are a distinct, segregated, and accessible part of the colony. Collecting them on baits can unambiguously identify them as foragers. Because they are a separate, territorial population, there is no immigration or emigration, and if mark loss can be minimized, forager "birth" and death can be dealt with by multiple marking events.

I used to demonstrate the principles of mark-recapture to my classes using a jar containing an unknown number of white beans into which I mixed a counted number of black beans. I then asked students to withdraw small random samples of beans and compute the percentage that were black. From 8 or 10 such samples, we calculated how many beans the population of beans (including both colors) contained. The estimate was usually remarkably close to the true number. The principle here, as in mark-recapture with animals, is one of dilution. You could as readily determine the volume of an unknown amount of liquid by mixing in a known amount of a solute (a dye would work well), followed by determining the solute concentration after complete mixing. In effect, you are counting the number of molecules recovered in the second samples. Say you needed to know the amount of water in your swimming pool. You could dissolve 1 g of the dye rhodamine B and the next day determine its concentration spectrophotometrically. Say it ended up being 0.000005 g per liter—well, there's your answer: 200,000 liters, the inverse of the concentration. Surely more accurate and less work than counting how many five-liter buckets it would take to empty the pool, and you have a nice pink pool when you are done.

Back to the estimation of forager populations: for many purposes, ants need be marked only as to group membership, so all you need is one color per group.

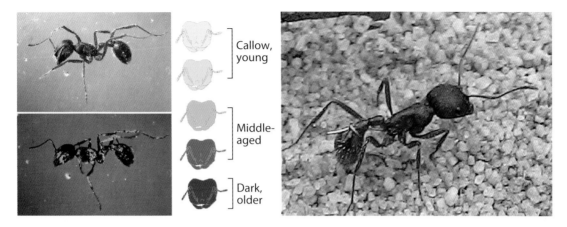

FIG. 8.3. *Left*, marked workers of *Pogonomyrmex badius* under visible and ultraviolet light. *Center*, worker age as judged from their cuticle color. *Right*, a worker with a wire belt. Author's photos.

In a mark-recapture study of foragers, you might want a different color for each marking event, but because of practical limitations, that is not likely to be more than four colors. As a master's student in the early 1980s, Sanford Porter developed a method for marking harvester ants using fluorescent printers' ink dissolved in ether. When he joined my laboratory as a PhD student, he brought a couple of his one-pound cans of printers' ink with him, enough ink to mark millions of ants over many years. It formed a tough skin in the jar, but by poking a hole in the skin, we could extract the fluid ink below. It was over 15 years before I had to buy another few cans in several fluorescent colors from Gans Ink Co.

The ink was dissolved in ether and sprayed from a perfume mister, and most of the ether evaporated during transit to leave tiny, sticky, invisible dots on the ants that hardened like paint and shone like beacons under ultraviolet light (fig. 8.3). The nonpolar ink and ether ensured that the ink stuck well to the ants' waxy cuticle. Christina tested the durability of such marks on *Pogonomyrmex badius* by tying fine copper wires around the waists of thousands of workers of a range of ages, as judged by their cuticle color using a scale (fig. 8.3), and spraying them with fluorescent ink before returning them to their colonies. When these workers, recognizable by their wire belts (fig. 8.3), appeared on the surface as foragers up to six months later, they all still bore the fluorescent ink mark. The ink was as good as the wire, and a whole lot less work. This made us confident that marking with ink alone was suitable for experiments that ran for weeks, months, or even years. Our top priority was to figure out how the superorganism allocates workers to the forager labor group, and how this allocation depends on colony size and season. To enlarge on the superorganism metaphor, we sought to determine the size and the size variation of one of the "organs" of the superorganism. Other "organs" could follow later. I must emphasize again that "organ" is not meant to be taken literally but is a shortcut way of building on the parallels between the subdivision of major life functions within an organism and a superorganism.

THE "ORGANS" OF THE SUPERORGANISM

The challenge thus boiled down to this: How much of a colony's total resources was assigned to finding and collecting food for the colony, and how did this vary with season and colony size? Christina set out to answer this question in several local ant species. Most species were problematic for one reason or another, but *P. badius* turned out to be a myrmecologist's dream: it was large bodied and had colonies of moderate size that were easy to find because the ants advertised them by covering the nest disk with charcoal. Moreover, we already knew a good deal about its natural history as a result of my studies in 1989–1990, as well as earlier work by Frank Golley and John Gentry in South Carolina.

Christina became a regular fixture out at Ant Heaven during every month of the year for three years, determining the proportion of workers that colonies allocated to foraging. On several occasions she had an audience in the form of a curious black bear that watched patiently from 50 m away, and one summer a gray fox curled up nearby and watched her work or licked up the cookie crumbs she used as ant bait.

Each determination required three days to complete. On day 1, Christina drizzled a circle of birdseed about 1.5 m from the nest entrance to entice foragers to do their job. Any ant that picked up a seed and headed back toward the nest qualified as a forager and was captured. Over three to six hours, she captured up to 93% and never less than 35% of the actively foraging population. After counting, she sprayed them with fluorescent printers' ink and released them back to their nest, where they went about their jobs as normal. The next day, she used crushed shortbread cookies instead of birdseed, because the ants seemed to get their fill of birdseed on day 1 and were no longer excited enough to collect it. Once again, she captured a sample of foragers, but this time she counted the foragers that bore a fluorescent mark under ultraviolet light. The proportion marked yielded an estimate of the number of foragers in the colony, because the number marked and released on day 1 was the same proportion of the total forager population as was the proportion marked in the sample on day 2. About 60% of foragers marked on day 1 were recaptured on day 2, imparting great confidence in the population estimates.

HOW LONG DO FORAGERS LIVE?

We first used this method to determine how long foragers live and how fast the forager population turns over. The foragers released on day 1 gradually die off, becoming an ever-smaller proportion of the recapture samples. But how can we tell whether this is because they die, or because unmarked workers become foragers, or both? We can distinguish between these choices by doing a second mark-recapture-release estimate a couple of weeks after the first one, using a different color ink. The second mark will reveal the number of foragers on the second date, and knowing both how many foragers were marked on the first date and how many still bear this mark on the second date, we can calculate how many of the first group are still alive. Application of this procedure to many colonies showed that once a worker began foraging, her average life expectancy was about three to four weeks (27 to 38 days—which, by the way, is also true of fire ant foragers). Thus, about 3%–5% of the forager population dies each day, and the entire forager population is completely replaced every month or so. The colony suffers these loss-and-replacement costs continuously during the foraging season but partly reduces the energy loss through the depletion of the fat contained in each worker by the time she becomes a forager.

WHAT PROPORTION OF THE COLONY FORAGES
THROUGHOUT THE YEAR?

When the purpose of the procedure was to determine the proportional invest-ment in foragers, we also needed to know the total colony size. On day 3, armed with a shovel, trowels, a vacuum, and many trays, Christina excavated and captured each colony in 20 cm depth increments and then counted all the adults, larvae, and pupae. Dividing the total adult worker population (includ-ing the foragers) into the number of foragers determined on day 2 yielded the percentage of the colony allocated to foraging. As we saw in chapter 3, exca-vating a colony is no small undertaking, as they are typically 2 to 3 m deep, and Christina is a scant 157 cm tall and therefore needed a short stepladder to get out of the hole. Over the three years of this part of the study, she exca-vated a total of 55 colonies, more evidence of the enormous effort it takes to establish a few simple scientific facts.

The first and most obvious finding was that foragers were never deeper than 12 cm in the nest. In other words, they had very little contact with life deeper in the nest, including brood and young workers. This will become important in the next section.

The onset of foraging preceded the first appearance of brood by a month and extended beyond successful brood rearing in the fall. The proportion of the colony allocated to foraging changed dramatically over this season (fig. 8.4), rising from zero during the winter (November to March) to a peak of about 35% for mature colonies in early July, and then declining back to zero by late November. The general pattern was similar every year. This pattern could have arisen by changes in the forager population, changes in the total nest popula-tion, or both. The answer to this lay in a parallel study of two groups of colo-nies, one in which the seasonal change in the proportion foraging was deter-mined, and the other in which the size of the forager population of each colony was determined several times throughout the season. The first group was dug up as above, but the second group was not. From the proportion foraging in the first group, the colony size of the second group could be calculated "back-

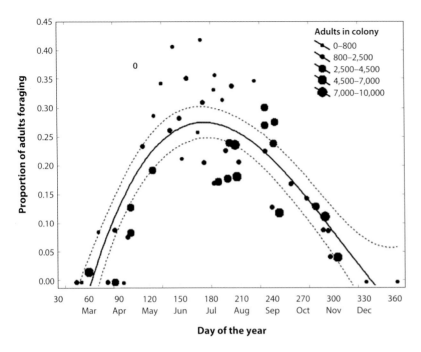

FIG. 8.4. Seasonal variation of the percentage of workers allocated to foraging. From Kwapich and Tschinkel (2013).

ward" from the size of its forager population. For example, if 25% of colonies in the first group were foragers and a colony of the second group had 1,000 foragers, then the size of the second colony was 4,000 workers.

In this way, Christina showed that the changing percentage of foragers was the result of changes in both the number foraging and the colony size during each season. In mature colonies, no new workers appeared until late June, following the completion of sexual production. This lack of new workers combined with the entry of older workers into the forager class and their death without replacement, leading to a decrease in colony size and an increase in the proportion foraging. This intense foraging thus coincided with sexual production and peaked in June and July. Once the sexuals were done, the birth of workers at high rates once again increased colony size, thus diluting the forager population from August to November. At the same time, dead foragers were replaced so that the number of foragers remained constant (although their proportion declined) until September or October. At that time, the replacement of dead foragers sharply decreased, so that by November there were no foragers anymore.

Colonies smaller than about 700 workers did not divert any production into sexual ants and were therefore able to allocate a higher proportion (about 40%) of their labor to foraging than did reproductive-sized colonies. This probably speeds colony growth in small colonies, but the seasonal pattern of allocation is the same for reproductive and nonreproductive colonies.

During the main brood-rearing season between May and October, the colony allocated a constant 1.6 foragers per larva in spite of the dramatic change in the number and percentage of foragers—in other words, foragers were coordinated with larvae. But was this a response to demand, or was it regulated in some other way? What would happen if the demand were increased or decreased? Would the forager population change accordingly?

HOW IS THE FORAGER POPULATION REGULATED?

There are two easy ways to change the ratio of demand to the number of foragers available to meet that demand. We can remove, say, half the foragers, or we can double the number of larvae by adding larvae from a "donor" colony (larvae are readily accepted between colonies). When Christina removed half the foragers and then did a second forager estimate a week later, she found that the foragers had not been replaced, and that the rate of new forager appearance was the same for both the "forager-removed" and the control colonies. Thus, the colonies had not moved workers into foraging in order to meet the increased demand per forager. Instead, when she excavated these "forager-removed" colonies, she found that half the expected larvae had disappeared and the forager-to-larva ratio was still 1.6, the same as in the control colonies. The colony had reduced its larval population to bring it into line with the available forager sector. The foragers were not replaced until younger workers simply aged into foragers.

Increasing demand by doubling the larval population (without changing the forager population) produced similar results—a week later, the forager population was unchanged compared to controls, but upon excavation, Christina found that half the larvae had disappeared, so that the forager-to-larva ratio was still 1.6. And this did not result from the rejection of the "foreign larvae," for these had been marked with fluorescent food, showing that the disappearance was unbiased—native and added larvae were both affected equally. No additional foragers had been recruited to service the added larvae. Instead, the excess larvae had been "culled," to use a euphemism.

In a third experiment, Christina removed half the foragers and kept them in the lab for 20 days, enough time for the stolen foragers to have been replaced by normal worker aging. When she added these workers back after 30 days, the forager population was boosted by about 40% and remained at this level for some weeks. The absence of any reversion to in-nest duties once again shows that becoming a forager is irreversible.

In all these experiments, the lack of response to demand (or supply) suggested that the population of foragers was regulated mainly by demography, meaning that the kind of work ants do is associated with their age on a fixed schedule. This schedule could not be accelerated by an increase in demand for forager services or reversed by an oversupply; it proceeded only at the deliberate pace of natural worker aging. This, then, is one of the gears of the superorganism machine.

But what if we could keep workers from dying at such a high rate, thereby increasing the forager-to-larva ratio and slowing the need for forager replacement? To do this, Christina penned colonies in screen-bottomed cages so workers could not go afield to forage and face an early death (fig. 8.5). Of course, she provided food in the cages so that whatever effects she found would not be attributable to starvation. Controls were caged-unfed colonies and breached-cage colonies able to forage afield. The forager population was estimated by mark-recapture just before penning, and again 20 days later.

Over the 20-day experimental period, foragers in caged colonies lived 57% longer than controls, and the time for complete replacement of the forager population increased from about 27 days to 55 days. This suggested that the causes of their deaths lay in the field, and they were not dying of old age. Indeed, workers

FIG. 8.5. A colony caged in a screen-bottomed enclosure to prevent foragers from going afield. Author's photo.

captured as foragers lived many months in the lab. Equally interesting, reducing forager death in the field did not change the size of the forager population but reduced the rate of entry of younger workers into the forager population from 69% to 43% in 20 days, a negative feedback in which the forager population inhibited the appearance of new foragers. As a result, the younger workers that the colony continually produced during the warm season were not being sent off to the slaughter as foragers, and the colony grew in size beyond what it normally would have if we had not reduced forager mortality by caging. Although foragers are the leanest workers in the colony, it is not the loss of fat that drives the transition to foraging—penned and fed foragers reverted to the fat content of middle-aged workers but were still recruitable as foragers when unpenned after 20 days. Foraging is not reversible.

The difference between longevity in the field and the lab suggests there is something out there that makes the foraging territory a killing field. Why is foraging so dangerous? Is it the physical stress of water loss, overheating, or getting lost, or is it predators, disease, or enemies? The next experiment was stimulated by the observation of occasional attacks by one colony on another, and by finding foragers with the heads of others firmly clamped to a leg or waist. So instead of caging the focal colony, Christina now caged all of its neighbors and once again determined forager life span, turnover, and population in the focal colony. Caging neighbors, like caging focal colonies, extended the life span of foragers about the same as caging the focal colonies, and it reduced their turnover, suggesting that foragers die at least in part by fighting with foragers from neighboring colonies. They are the soldiers, the cannon fodder in the continual border skirmishes that allow colonies to maintain the space and resources they need to live and flourish.

TWO KINDS OF FORAGERS

We thus have strong evidence that foragers are allocated mainly by demography on a fixed schedule of aging, not demand. But it gets much more complicated—if foraging and brood production are limited to the warm season during which dead foragers are continually replaced, and the last young workers of the year are produced before November, then the spring foragers must have spent the winter in the nest. Christina wire-banded the callow workers of several colonies throughout the year and reestablished them in ice nests. When they were old enough, these banded workers appeared on the surface as foragers. Workers from the June to August brood were about 40 days old when they appeared, but workers in the late summer and autumn brood did not appear until spring, when they were over 200 days old. Once they became foragers, both groups had a life expectancy of three to four weeks because death came from extrinsic causes, not old age (see above). These two cohorts of workers thus differed greatly in the rate at which they aged physiologically and behaviorally. The spring workers darkened rapidly and galloped through their age-related changes of behavior to become foragers in only 40 days, whereas the fall workers did not forage until late spring, when they were over 200 days old and were often still recognizably lighter in color. Clearly, the rate of aging can be modified by evolution, just as life span can.

We can now see how differences in demography control the foraging population and limit it to the warm, brood-rearing season. The slow-aging workers born in the second half of the foraging season do not age into foraging until the next spring, gradually boosting the foraging population early in the season. Because they die within two to four weeks of becoming foragers in the spring, the slow-aging worker population is gradually depleted. However, once sexual production is underway, the colony produces fast-aging workers that emerge after the sexuals and begin to forage at only 40 days of age, so that they forage side by side with the declining population of slow-aging foragers. Eventually, all foragers are of the fast-aging type. This replacement continues until about September, when the colony switches from producing fast-aging workers to slow-aging ones that do not reach forager age until the following spring. Therefore, in late summer, there are no longer any fast-aging (or any) workers to replace foragers that die, and the forager population dwindles to zero by early November. This alternation of workers with different demographies thus drives the seasonal foraging patterns.

This simple seasonal control system leaves us with this question: How does the colony control the rate at which workers age? What physiological mechanisms bring about this dichotomous and extreme rate of aging? If aging rate is so readily malleable even within the life cycle of one species, then clearly natural selection modifies it across species as well.

These demographic differences probably also drive the other labor transitions that take place underground and out of sight because we can expect slowed-down aging to affect all these transitions. Thus, all activities are retarded until early spring when the youngest slow-aging workers begin brood care, rearing sexuals and then workers, and the oldest begin foraging.

CONNECTING FORAGERS WITH BROOD-CARE WORKERS

We now know quite a lot about how much this particular superorganism invests in one functional group, the foragers, how this varies with colony size and season, and how this labor group is regulated. Could we do similar estimates of allocation to other major labor groups? If foragers never venture deeper than 15 cm in the nest, and the brood and their caretakers spend all their time deep in the nest, this raises the question of how foraged food reaches the brood 1 to 2 m meters below, and how seeds reach the storage chambers 30 to 80 cm deep. How do trash and excavated sand from deep in the nest get transported to the surface? All this suggests that there must be a group of workers that shuttle between these two noncontacting groups. But how can we demonstrate the existence of a group whose actions all occur belowground?

The key lies in the workers in the uppermost chambers. Even when almost all the foragers coming to baits were marked, as few as half the workers in the top 15 cm of the nest were marked. In other words, in addition to foragers, the top region of the nest was occupied by workers that did not forage and thus escaped marking, but shared chambers with foragers. These were prime candidates for a group of workers, distinct from foragers and brood-care workers, that shuttled seeds and food brought in by foragers to the deep nest regions where the brood and their caretakers waited. These same workers probably shuttled excavated sand and trash upward for disposal. Let us call them transfer workers. So if we could simply mark the unmarked workers in this top region with a different color than the foragers, we would have tagged this candidate group.

The first step was thus to exhaustively mark all the foragers by repeatedly baiting colonies over a few days until 75% to 92% of the foragers were marked. As simple as this plan appeared, the problem was that in order to capture these putative transfer workers, we had to dig up and destroy the upper nest region. My assistant Nicholas Hanley and I exposed each of the three to four chambers in this top region to capture all the workers and trace the outlines of each chamber on transparent acetate. We then cut out a facsimile of each chamber in a sheet of Styrofoam insulation (fig. 8.6). Fastening a roof of acetate on each, we buried each at its original depth and orientation and provided a shaft to connect it to the chamber above and below. It was a great relief when the ants seemed to accept these ersatz chambers as though they were the original. They hardly seemed to notice. Over the course of two summers, Nicholas and I applied this procedure to 11 colonies.

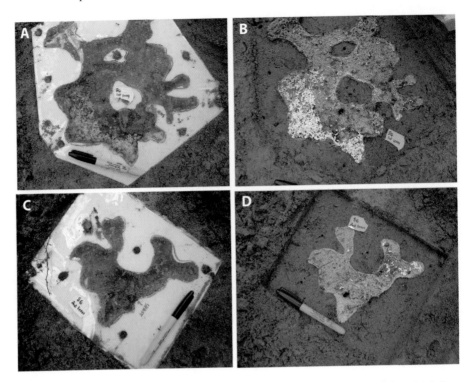

FIG. 8.6. After the upper chambers of *Pogonomyrmex badius* nests were raided (and destroyed) for workers in the transfer-worker experiment, they were rebuilt in their original shape and location using cut-out forms of Styrofoam (*A* and *C*). These chambers were exposed after a few days and showed that the ants accepted and used them readily (*B* and *D*). The surrounding soil has been digitally darkened to make the chambers more visible. Author's photos, from Tschinkel and Hanley (2017).

Having captured all the workers present in the top nest region, we separated the marked foragers from the unmarked workers. If the foragers had been marked green, the unmarked workers were now marked orange. Although a few of the orange workers were really foragers, the vast majority were not. All of these marked workers were then returned to the reconstructed nest, where they happily went about their quotidian tasks—the foragers returned to foraging and the others returned to their subterranean duties.

Two to six days later, after colony life had settled into what we assumed was the old routine, we excavated the entire nest chamber by chamber (see chap-

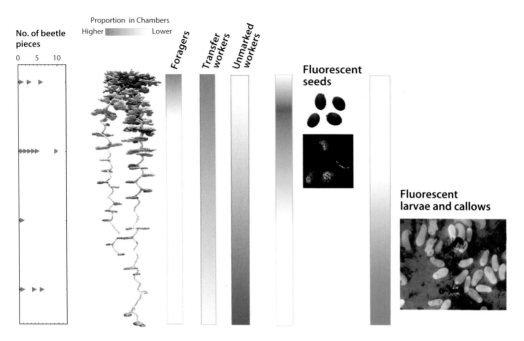

No. of beetle pieces

0 5 10

Proportion in Chambers
Higher   Lower

Foragers

Transfer workers

Unmarked workers

Fluorescent seeds

Fluorescent larvae and callows

FIG. 8.7. Foragers rarely go deeper than 20 cm in the nest, but transfer workers roam the entire vertical nest, moving seeds, soil, food, and probably brood. The proportion of each type in chambers is represented by the darkness of the shading in the columns. Transfer workers quickly move marked seeds (*red column*) and fluorescent-dyed food downward (*left graph*). Larvae that eat this marked food fluoresce brightly under ultraviolet light (*green column* and *right photo*). Unmarked workers are probably mostly involved in brood care and are thus largely coincident with larvae.

ter 3) and kept the contents of each chamber separate for inspection for marked workers under ultraviolet light. This showed the distribution of forager and transfer workers from the surface to the deepest chamber of the nest. As before, foragers were limited to the surface and the top 15 cm of the nest. The very few found deeper were probably foragers returning from the field that fell undetected into the pit during our excavation.

On the other hand, almost a third of the transfer workers that had been initially captured in the top 20 cm of the nest were now below this depth, all the way to the bottom of the nest (110 to 120 cm) (fig. 8.7). They were particularly abundant down to the normal seed chamber depth (70 cm), forming 10% to

35% of workers in chambers, but even below 70 cm they averaged 6% (maximum 10%) of the workers in chambers. Young workers rarely left the deeper regions of the nest and were thus unmarked.

For the later replicates of this experiment, we added another refinement by offering each colony fluorescent-marked seeds and fluorescent-dyed pieces of mealworms an hour or two before we commenced the excavation. Foragers readily delivered these items into the nest and dropped them in the uppermost chambers. The idea was to give transfer workers something to do, and finding the dyed items below 15 cm would show that they had quickly sprung into action and transported these items downward. Because foragers never ventured deeper than 20 cm, any transport had to have been carried out by the transfer workers, for these were the only other workers with whom the foragers shared space. This strategy yielded great results—not only did we find fluorescent-marked seeds down to 50 cm and mealworm pieces down to the brood chambers, but many larvae had fed on the mealworms and now glowed like lanterns under ultraviolet light (fig. 8.7, *right photo*). The transfer workers had accomplished all this in one to three hours. Having finished inspecting all the nest contents, we made an ice nest (see chapter 5) and returned each colony to its own territory. They all moved in immediately, allowing us to track their fates for a few weeks.

THE TRANSFER WORKERS ARE AN AGE GROUP

The best way to understand these results is to suppose that there is a distinct group of workers that shuttles between two noncontacting groups—foragers in the top of the nest and brood and their caretakers in the deep regions. In most ant species, the foragers are the oldest workers, and it now seemed likely from the location of the transfer workers that they were the next younger workers and would become foragers as they aged. And sure enough, capturing foragers up to a month later revealed that workers marked as foragers had disappeared from the population (i.e., died), and workers marked as transfer workers had appeared as foragers. Thus, younger workers are more or less restricted to the lower nest regions and are not very mobile. As they age, their vertical mobility increases and they act as transfer workers. Eventually, they roam up and down most of the nest and into the top chambers, eventually probably acting as trash workers or sand dumpers on the surface before they gain the necessary skills and orientation capacity to become foragers in the last two or three weeks of their lives. Because workers are born in the bottom of the nest and move upward as they age, there is a continuous conveyer belt of replacements, first to shuttle materials up and down, and finally to scour the killing fields that are the source of the colony's food. The workers' changing duties are linked to both age and location in the nest. This age-related movement and change of location are the gears of the machine that is the superorganism (fig. 8.8). Perhaps it is stretching the organ analogy too far, but functionally, foragers correspond to the feeding/hunting/foraging behavior of unitary animals, transfer workers correspond to a circulatory/transport system, and brood-care workers and brood correspond to mitotic growth. All these functions are necessary parts of the superorganism as much as they are of unitary animals.

OVERVIEW OF A SUPERORGANISM

We have now dissected the colony into three major "organs," each carrying out a major colony function—seasonal foragers that find and retrieve food and defend territory; transfer workers that move material within the nest as well as store and retrieve seeds; and by subtraction, the unmarked brood-care workers responsible for feeding and grooming larvae and the queen. None of these groups perform only a single task; rather they carry out sets of related tasks that define broad needs of the superorganism. Moreover, there may be age-based transitions even within these major groups. Thus, the transfer workers that dump sand on the mound surface are more likely to become foragers than other transfer workers. The flexibility of worker behavior within these major task groups is still an open question. There is probably flexibility in what particular items a transfer worker transports—sand, seeds, food, or larvae. There might also be flexibility in what part of the nest constitutes a transfer worker's

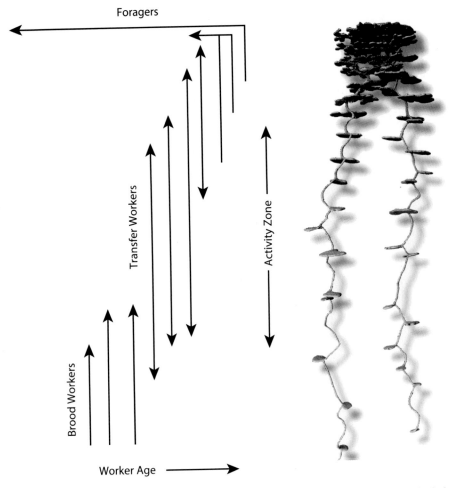

Fig. 8.8. Schematic of age-related upward movement and changing behavior. From Tschinkel and Hanley (2017).

"beat"—surface to seed chambers, seed chambers to brood chambers, or the whole nest. The gradual entry of "sand workers" into the forager class supports the generally upward movement of aging workers, possibly creating limited "beats." I have not yet figured out how to do removal experiments with transfer workers to test transfer worker flexibility or the existence of limited beats.

Similar questions exist for brood-care workers—how flexible or reversible is their transition to transfer workers? The sharpness and irreversibility of the transition to foraging is not necessarily echoed in the other transitions. There are good reasons to limit forager contact with other nest members, as this reduces the transfer of contagion. And once a worker has gained enough skills to forage effectively, she is much more valuable to the colony in this role than in less-skilled roles within the nest—it doesn't take an ant genius to carry sand pellets or seeds. There are also good reasons for not replacing foragers on simple demand, for should something in the field cause extreme mortality, the whole colony could be catastrophically depleted. Better to forgo foraging and munch stored seeds until enough workers age into the foraging class. In western species of *Pogonomyrmex*, when a horned lizard parks itself at the nest entrance and gulps down one worker after another, the colony stops foraging. Some have suggested that the colony has "figured out" the threat and responded, but it seems more likely that the horned lizard has exhausted the supply of foragers.

Mark Twain wrote, "There is something fascinating about science. One gets such a wholesome return of conjecture for such a trifling investment of fact." In this spirit, I want to indulge in a speculation: if the up-and-out trend of workers continues without limits, foragers should venture farther and farther from the nest as they age, possibly ultimately encountering workers from neighboring colonies, getting into fights, and dying. The distance from the nest where this occurs will determine the size of the colony's foraging territory, which in turn will determine available resources, which in turn will feed back on colony size and sexual production. The colony that fields more foragers will encounter neighbor foragers farther from its own nest and closer to the neighbor's, thus claiming a larger territory. Such border skirmishes may be why penning neighboring colonies extends the lives of foragers in the focal colony. I will leave it for someone else to test this interesting hypothesis.

DEATH COMES TO THE SUPERORGANISM

Like unitary organisms, superorganisms have a characteristic average life span, along with a chance of dying that changes with age. For most single-queen colonies, queen life span is probably synonymous with colony life span. For a range of reasons, colony life spans determined in the laboratory should be taken with a grain of salt, but sadly, colony life spans are rarely estimated in the field. To determine life span in the field can be a very long project if one tracks a set of marked colonies from birth to death. However, there are shortcuts that could reduce the required time to a few years. The first of these is based on the rate of population turnover—that is, the percentage of the population that die and are born each year. If these two rates are about equal (and in a stable population, they are), the reciprocal of the annual turnover is the average life span. Let's say that 10% of colonies die each year and are replaced by new colonies. The average life span is then the reciprocal of 0.1, or 10 years. My postdoc Eldridge Adams and I tracked a population of 1,000 fire ant colonies for six years and found a turnover of 13%, which yielded an average life span of 8 years. A second, more sophisticated method is based on the number that die each year, yielding the probability of dying—that is, the mortality rate. At the end of the study, survivors are "censored" from the analysis because they give no information on the probability of dying. This method also yields plots of the chance of still being alive as time progresses. Such plots can also be produced for subgroups within the population, for example by sex, age, or size.

The data for this analysis came from the population of 400 individually tracked harvester ant colonies (see chapter 4). For six years, my small army of helpers and I visited each colony four to seven times per year to see whether it was still alive and whether it had moved. A colony that showed no activity for a year was declared dead. If it had moved, we moved its tag to the new location. If it was a "newborn" colony, we gave it its own number. At each visit, we also measured the diameter of its nest disk, for this correlated very strongly with both the number of ants and the volume of the nest and allowed us to sort colonies into size classes for the computation of average mortality rates.

The results showed why it is important for colonies to grow as rapidly and as large as possible (fig. 8.9). Twenty-five percent of colonies in the smallest size class died each year, yielding an average life expectancy of four years. As colony size increased to midsize, the probability of dying decreased to about 6% per year and the life expectancy increased to 17 years. Over 90% of colonies in the largest size class were still alive after six years, giving them a life

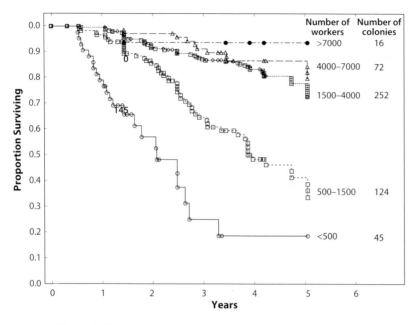

	Number of workers	Number of colonies
	>7000	16
	4000–7000	72
	1500–4000	252
	500–1500	124
	<500	45

FIG. 8.9. The probability of survival is strongly dependent on colony size. The smallest size class has a life expectancy of about 4 years, and the largest over 30 years. The curves show the probability of still being alive in *n* years for colonies of different initial sizes. From Tschinkel (2017b).

expectancy exceeding 30 years. Interestingly, these rates depended primarily on colony size, rather than age. Many colonies never attain extreme sizes despite being older, and they die at the rate characteristic of their size, not their age. There is thus safety in size. Failure to reach a "safe size" could be the result of landing in a poor site or a tough neighborhood, or of unfavorable environmental conditions such as drought, excessive rain, failed seed production, an unfavorable genotype, poor mate choice, and so on. Colonies with more and larger neighbors also had significantly shorter life spans, perhaps because attaining a "safe size" was more difficult for them.

The visible evidence of colony longevity is that as I write this in 2019, many of the bigger colonies first labeled in 2010 are still alive. I fully expect that were I to survive until 2040 and be able to stagger around Ant Heaven as a centenarian, I would still see a few of "my" colonies thriving. What a nice 100th birthday present that would be!

THE VALUE OF A LONG LIFE

Longer life span suggests more opportunities for reproduction. My student Chris Smith dug up and censused 19 harvester ant colonies during the spring and summer in order to establish the relationship between colony size and sexual production (as with most of my students, his project involved digging holes). Chris found that the production of sexual ants is directly proportional to colony size—doubling colony size means more or less doubling sexual production. This means that there is a second, equally important benefit to large size—all other things being equal, the chance of establishing or producing a daughter colony is the product of annual sexual production and life span. Relative to the fitness (lifetime sexual production) of the smallest colonies that produce sexuals (~700 workers), the fitness of colonies of about 2,000 workers is 40 times as large, that of 4,000-worker colonies is 150 times as great, and that of colonies with over 6,000 workers is 350 times as great. In reality, colonies grow, thereby reducing but not eliminating the fitness differences among them. However, at Ant Heaven, the great majority of colonies do not grow to extreme size but vacillate around an average size of 2,000 to 4,000. In 2014, only 17 of 400 colonies were larger than 6,000 workers. No wonder colonies compete for the resources and space to grow!

A PANORAMIC COMPARISON WITH UNITARY ANIMALS

Our representative seasonal superorganism, the Florida harvester ant colony, is composed of labor groups that are the analogues of the organs of unitary animals, carrying out major life functions and changing with the seasons. As in unitary animals, seasonality results from the need for creatures to acquire energy faster than they use it. When this is not possible, such as during winter, various adaptations reduce energy use until conditions once again become favorable. Similar considerations apply to reproduction—some seasons are better than others.

With these principles in mind, we see that in the fall, harvester ant colonies prepare for a hibernation-like state in ways that echo what unitary animals do—

both shut down reproduction, store fat and protein reserves, and then cease foraging. Unitary animals do this by gonad regression and changes in behavior, and colonies by shutting down the queen's gonads and not replacing foragers that die. Animals reduce their metabolism to save energy and slow the aging of their bodies. In colonies, overwintering workers age much more slowly than summer workers. We do not know whether their metabolism slows down, but I would wager that it does. During the winter inactive period, animals metabolize their fat stores for maintenance energy, and colonies live on both copious seed stores and the fat stored in the bodies of young workers. These are similar solutions to similar problems, one at the organism level and the other at the superorganism level.

For many temperate animals, spring is the optimal season for reproduction—it synchronizes timing across the population and allows enough time for offspring to grow to a size that can survive the winter. In many, it may also allow multiple reproductive attempts. Because foraging is often meager in early spring, many animals begin reproduction with metabolic stores acquired the previous year. We see the same pattern in our harvester ant colony—the production of sexual ants, the agents of colony reproduction, begins before the season is favorable for foraging and is fueled by fat stored in the bodies of young workers the previous fall. The rate at which these workers become lean and age into foragers is coordinated with larval demand for food through demography.

What animals do after they have reproduced varies greatly depending on their life histories. Some invest a great deal of time and energy in parental care, while others may reproduce again, but both require a great deal of energy. Eventually they must all prepare for the next winter, ending reproduction and feeding to put on metabolic stores. Our harvester ant colony produces only one brood of sexuals each year, and because most of these depart on mating flights between late June and early July, there is no parental care beyond providing the female sexuals with body reserves. By the time sexuals have flown, colonies have expended a large fraction of their overwintering workers, thereby losing colony size much like unitary animals may lose body weight during reproduction. Having reproduced, both unitary animals and colonies regain size and if possible increase it, the former by growing, the latter by producing large numbers of rapidly aging workers that build up colony size and metabolic stores. Just as body size is positively related to reproductive output in most animals, colony size is similarly related to sexual production in ant colonies (see above). Achieving large size serves fitness in both.

The beginning of life is also governed by parallel rules. Young unitary animals invest heavily in growth and reproduce only after reaching a threshold body size. Similarly, newly founded ant colonies invest heavily in worker production so that colonies grow rapidly. Only after the colony reaches a threshold size does it produce sexual ants, the agents of colony reproduction. For both, the age at first reproduction strongly affects population growth, while the probability of successful reproduction selects for life span.

I have now shared what I have learned about the architecture of ant nests, its variation across species, the labor that creates the nest, and the effect ants have on soils as a result of all this digging. All of this variety and complexity is the product of millions of years of evolution. Modern ants, their behavior, and the architecture of their nests arose through the action of natural selection on the ancestral ant and her behavior. Because her descendants can be arranged into a family tree that reflects their degree of relatedness as well as how long ago their lines of descent diverged, a reasonable question is whether the architecture of their nests can also be arranged into family trees. Are such trees parallel to those of the ants constructing the nests, or is nest-digging behavior so volatile that there is little parallelism? Attempting such an architectural family tree, as I do in the next chapter, might at least illuminate the changes in digging behavior that led to the diversity of modern ant nest architecture.

CHAPTER 9

The Evolution of Nest Building by Ants

THE ANCESTRAL ANT NEST

The ancestor of all modern ants dug the ancestor of all modern ant nests. What did this nest from which all others descended look like? No fossils of either this ancestor or this nest are known to exist, so we must reconstruct both through indirect evidence and reasoning, using comparative morphology. This evolutionary reasoning starts with a diverse set of modern structures and reasons backward to what the original ancestral structure must have been. Underlying this method is the principle of homology—the idea that it is possible to identify structures that descended or evolved from a common ancestral structure. Such structures are called homologues (adj. homologous). Homology does not require that the modern structures perform similar functions or even that they look similar, only that they can be traced back to a common ancestral structure. For example, the wings of all modern insects are homologous, even though they may look very different and perform different functions. It seems likely that the ancestral insect had two similar pairs of simple wings, a conclusion that is confirmed by both the fossils of ancient insects and the near ubiquity of two pairs of structures in modern insects. Some modern insects have two pairs of very similar wings (unchanged much from the ancestral condition), while in others one pair is greatly modified into, for example, the wing covers of beetles (protective function, not flight), the tiny balance organs (halteres) of flies, or their complete loss in fleas.

These same principles of comparative morphology can be applied to other evolving biological entities, for example behaviors or social structure, or in our

case, ant nest architecture (or more correctly, the behavior leading to an excavated nest). Because the immediate ancestor of the ants was a solitary carnivorous wasp that nested in the ground, its nest was probably small and contained only the offspring of that mother wasp. This in turn suggests that
ancestral ant societies and their nests were also small. The vertical or nearly
vertical shaft is almost ubiquitous among modern ant nests, suggesting that a
vertical shaft was part of the first ant nest. Similarly, most modern ant nests
contain at least one or more chambers, meaning that chambers are a primitive
character, too. So we now have our ancestral ant nest—a small nest with a
vertical shaft perhaps 15 to 20 cm deep, and one or a few chambers (fig. 9.1).
The importance of this insight is that we can now view all modern ant nests in

FIG. 9.1. The ancestral ant nest probably consisted of a simple
shaft with one or more simple lateral chambers, as in this
reconstruction. Author's photo.

the light of how they evolved from this ancestral nest, and we can thus identify the evolutionary trends in nest architecture (or more exactly, the behavior of the worker ants that dig the nest).

EVOLUTIONARY TRENDS IN ARCHITECTURE

As ants evolved larger and more complex societies, chambers and shafts did not remain in their ancestral size and complexity. First and foremost, the modular nature of both ant society and its nest allowed the smoothest and simplest route to nest enlargement. Does a bigger colony have an evolutionary advantage? Just add more modules (workers). Does the now-larger colony need a bigger nest? Just add more modules (nest units, chambers, shafts). The principle becomes strikingly obvious when comparing the simple nests of the tuberculate fungus-gardening ant (*Trachymyrmex septentrionalis*, fig. 7.7) with the huge nests of the leafcutter ant (*Atta laevigata*, (fig. 9.2). The latter evolved through the multiplication of modules (shafts and chambers) seen in the *Trachymyrmex* nest until they accommodated colonies of millions of workers. The basic plans for both colony and nest seem almost infinitely expandable because they are modular. This principle can be seen in many ant species, especially those with very large colonies. It also operates in the evolution of plants—bushes evolve into trees through a similar multiplication of modules (leaves and stems).

However, nests grow not only through the addition of modules but also through modification of these modules. Thus, colonies created larger nests that differed from the ancestral nest in many details of shaft, chamber structure, relative size, and arrangement of chambers along the shaft(s). Many of these "changes of state" seem to be independent of one another, with various combinations leading to the range of nest architectures we see today.

We can assume that chambers in the ancestral ant nest were small and that they were excavated to one side of the shaft, because that is the arrangement in most modern ant nests. To judge from the range of chamber shapes found in modern ant nests, these small chambers were enlarged during evolution and growth through a number of somewhat distinct patterns in the allocation of

FIG. 9.2. The enormous nests of the higher fungus-gardening ants are examples of the increase in size by the addition of modules (chambers and shafts). *A*, excavation of a cement cast of *Atta laevigata* in Botucatu, Brazil, by Luiz Forti, Flavio Roces, and their coworkers (photo by Wolfgang Thaler). *B*, the surface mound of *Atta vollenweideri* in Formosa, Argentina. The wind-capturing turrets the ants construct to ventilate the nest are clearly visible (photo by Christoph Kleineidam). *C*, a detail of the *A. laevigata* excavation, with Flavio Roces for scale. *D*, cast and excavation of the nest of the grass-cutting fungus gardener, *Atta capiguara*, by Luiz Forti. Note the more dispersed location of fungus chambers and the longer connecting and foraging tunnels.

labor to the chamber walls that were being actively excavated (the "mine face") (fig. 9.2). Each chamber is a sort of fossil record of its labor "history," with a different history for long, narrow, broadly lobed, branching, round, and oval chambers. Chambers are enlarged horizontally away from their connections with the shaft, suggesting that workers enter a chamber, walk more or less toward a far wall, and get to work. The distribution of these excavators along the mine face determines whether the chamber is extended radially or in lobes that may be narrow or broad (examples in figs. 9.3 and 9.4). When some over-enthusiastic workers start digging at new spots, the lobes change from simple to branching. When the effort is three-dimensional rather than two, ovoid chambers result, like pearls on a string as in the northern tuberculate fungus ant, *Trachymyrmex septentrionalis* (fig. 9.4). A large colony of the leafcutter ant, *Atta* spp., may contain hundreds of such chambers, connected by angled or vertical shafts (fig. 9.2).

Local concentration of digging labor could result from some attraction between actively mining workers and new arrivals, a sort of desire to "be where the action is," or from certain geometric arrangements such as a shaft that connects at an angle and sends workers to the opposite side of the chamber, leading to a teardrop-shaped chamber. Large round chambers are relatively rare because they require workers to spread their efforts evenly over the entire

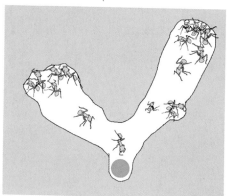
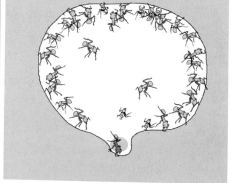

FIG. 9.3. Different distributions of worker digging effort result in different chamber shapes.

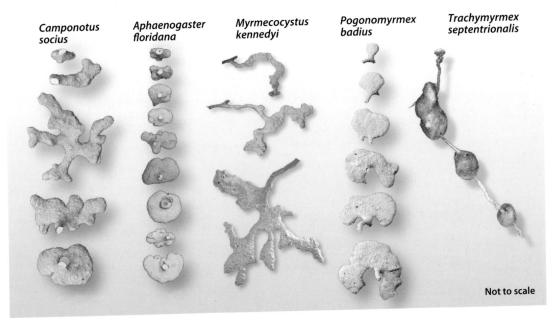

Camponotus socius **Aphaenogaster floridana** **Myrmecocystus kennedyi** **Pogonomyrmex badius** **Trachymyrmex septentrionalis**

Not to scale

FIG. 9.4. Chambers often change shape as they are enlarged, but the change differs across species, and even by depth within some species. Author's photos, in part from Tschinkel (2015a).

chamber perimeter, but this happens primarily when the vertical shaft is connected to the center of the chamber roof, making it equally likely for workers to go in any radial direction before getting to work (fig. 9.4).

In many species, the uppermost chambers just beneath the ground surface are different in their basic structure than deeper chambers in that they are formed of horizontal tunnels (I use "tunnel" and "shaft" in the sense that miners do, the former a horizontal and the latter a vertical tube). Rather than expand the walls of a growing chamber outward, they take their form from being driven forward by excavation only at the end of the tunnel without much lateral widening. Occasionally, a worker with a bit of rebel in her starts a side tunnel, but this too is driven only forward. Because these tunnels are not particularly straight and because they branch now and then, the outcome is a complex fabric of dividing and rejoining tunnels, or a starlike chamber (fig. 9.5). Usually, such structures are limited to the uppermost chambers—deeper chambers lose this complexity to become "ordinary" chambers (although there

FIG. 9.5. Subsurface chambers are often different from deeper chambers. *Top row*: small *Pogonymyrmex badius* (*left*) and a much larger one (*right*). *Middle row*: *Myrmecocystus navajo* (*left*), *Pogonomyrmex californicus* (*right*). *Bottom left*: *Dorymyrmex bureni*. The images are not to the same scale. Author's photos, in part from Tschinkel (2015a).

are a few exceptions). This raises the question of whether "tunnels" and chambers, although both horizontal, are produced by different behavioral programs, like chambers and shafts. Perhaps tunnels are horizontal shafts, expressed differently depending on depth, or perhaps they are narrow chambers. The answer will someday earn a fine PhD for an ambitious and imaginative student.

The height of the chamber is necessarily tailored to the size of the ant, but there is a lot of leeway in this, with large differences in height-to-area ratios. Workers of *P. badius* weigh four to eight times as much as those of *Aphaenogaster floridana*, and their chambers are much larger in area, but both are about 1 cm from floor to ceiling, giving *P. badius* a much larger area-to-height ratio.

Not to scale

FIG. 9.6. Extreme differences in the height-to-area ratio. The very thin chambers of *Pheidole barbata* (*left*) have a large area per unit height, while the dumpy chambers of *Formica dolosa* (*right*) have little area per unit height. Author's photos.

The large workers of *Formica dolosa* build thick, chunky chambers with a practical but inelegant appearance (fig. 9.6) and a low ratio of area to height. On the other hand, the extensive, perfectly flat, horizontal chambers of the tiny desert-dwelling *Pheidole barbata* are only 2 to 3 mm thick, with an enormous ratio of area to height (fig. 9.6).

These few examples of how chamber shape and size have changed from the ancestral condition, along with the suggested mechanism for these changes, allow us to provide a schematic for chamber evolution and enlargement. As before, I must emphasize that the evolving entity is not the chamber but the behavioral program that guides the workers in their excavations. It is not hard to see how the range of chamber shapes in figure 9.7 resulted from variation in the allocation of work. Perhaps for practical reasons, horizontality seems to be a largely conserved chamber character, as is approximate verticality of shafts. Ants respond to gravity as readily as people do.

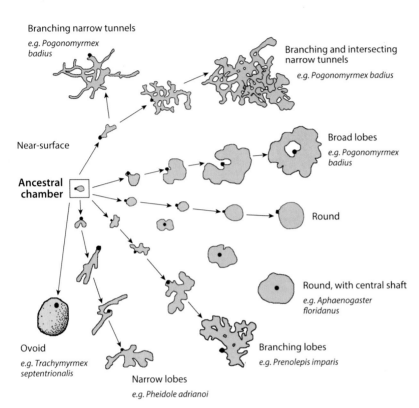

Branching narrow tunnels
e.g. Pogonomyrmex badius

Branching and intersecting narrow tunnels
e.g. Pogonomyrmex badius

Near-surface

Broad lobes
e.g. Pogonomyrmex badius

Ancestral chamber

Round

Round, with central shaft
e.g. Aphaenogaster floridanus

Ovoid
e.g. Trachymyrmex septentrionalis

Branching lobes
e.g. Prenolepis imparis

Narrow lobes
e.g. Pheidole adrianoi

FIG. 9.7. The changes in chamber excavation that led to a range of modern ant nest chambers. Such changes led to differences in chamber shape in the evolution and development of modern ant nests.

The arrangement of chambers along the shaft also undergoes large changes of state. As with chambers, these patterns are deduced from the range of variation found in modern ant nests, but the variation we see today did not necessarily arise in a regular sequence of evolution or development. Chamber spacing along the vertical shaft is highly variable and may be close, distant, irregular, or dependent on depth. Figure 9.8 shows examples of exceptionally close spacing (*Monomorium viridum*), exceptionally distant spacing (*Dorymyrmex bossutus*), and spacing that increases with depth (*Pogonomyrmex badius*).

However, some trends seem to be very strong and have clearly arisen independently a number of times. Among these is the relative concentration of chamber area near the surface, from both the creation of larger chambers there (fig. 9.5) and their closer spacing (fig. 9.8). This trend is necessarily accompanied

Not to scale

Fɪɢ. 9.8. Chamber spacing varies widely, with exceptionally close spacing in *Monomorium viridum* (*left*), exceptionally distant spacing in *Dorymyrmex bossutus* (*center*), and spacing that increases with depth in *Pogonomyrmex badius* (*right*). Author's photos, in part from Tschinkel (2015a).

by deeper chambers spaced relatively farther apart. Deeper chambers spaced more closely than shallower ones are rare. Very regular spacing occurs in some species (fig. 9.8, *Dorymyrmex bossutus*) but is also not very common.

Whereas shafts are predominantly vertical or nearly so, in some species the shaft is always angled (fig. 9.9). In a number of species, the shaft is angled in

FIG. 9.9. Although the shaft is predominantly vertical in most species, it is always angled in some, as in *Veromessor pergandei* from the Anza-Borrego Desert in California. Author's photo, modified from Tschinkel (2015a).

the upper region but changes to vertical at greater depth. This seems more common in desert species, but my sample is small. Distinctly helical shafts have so far been found only in the Florida harvester ant, *Pogonomyrmex badius* (figs. 2.2, 4.10), as described in chapter 4. In some species, shafts may branch and branch again, with each branch carrying chambers. Such branching is more or less distinct from the addition of entire shish-kebab units—that is, chamber-and-shaft units that extend downward more or less from the surface (fig. 9.10). Such multiple shish-kebab units are the major mode of nest enlargement in several species. In some species, such as the fire ants *Solenopsis invicta* and *S. geminata*, these units are packed so closely together that many of the chambers coalesce (fig. 9.10). In *Pheidole morrisi* and *Pogonomyrmex badius*, they are separate and distinct. One or more of the shish-kebab units is often shorter, suggesting that it was added later and is still being deepened. In *Ph. morrisi*, new shish-kebab shafts sometimes lack chambers, and this occurs occasionally even in *P. badius*. The condition that probably precedes such fused, multiple shish-kebab nests can be

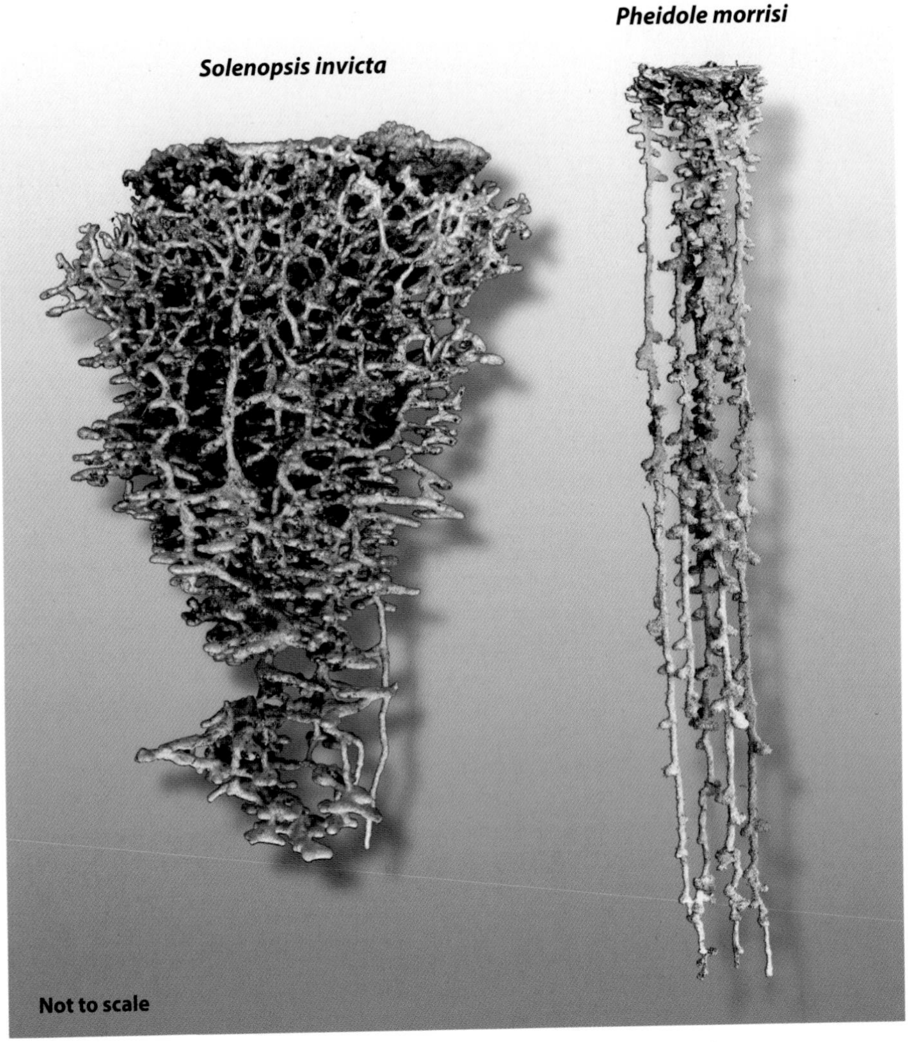

Pheidole morrisi

Solenopsis invicta

Not to scale

Fig. 9.10. In a number of species, nests are enlarged by adding new "shish-kebab" units, as in the fire ant *Solenopsis invicta* (*left*) and *Pheidole morrisi* (*right*). Author's photos, modified from Tschinkel (2015a).

seen in the enormous nests of the Australian meat ant (*Iridomyrmex purpureus*). This ant forms large clusters of closely spaced independent nests of the shish-kebab type, each nest with its own entrance and mostly without subterranean connections to neighboring nests (fig. 9.11). With the evolution of reduced spacing between such shish-kebab units, we arrive at nest architectures similar to

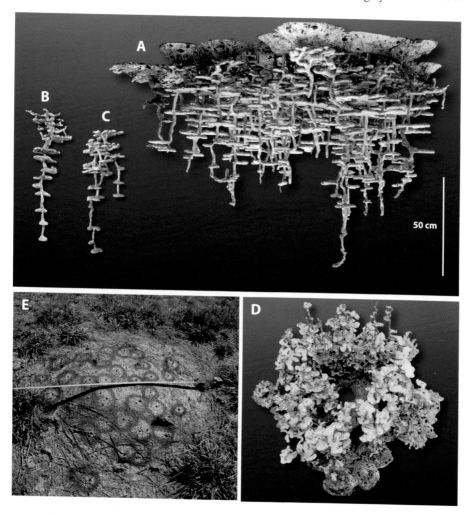

Fig. 9.11. The Australian meat ant, *Iridomyrmex purpureus*, forms clusters of relatively simple nests that are rarely connected belowground. *A*, a single cast of several dozen neighboring nests. The deepest shafts are over a meter deep, but many are not completely cast. The conical structures at the top are artifacts of casting that hold the cast together. *B and C*, two examples of single nest units. *D*, the cast viewed from below. *E*, the cluster of nests before casting, with all the independent entrances circled in red. All photos by Australian Ant Art, courtesy of Christopher and Stephen East, modified by Walter R. Tschinkel.

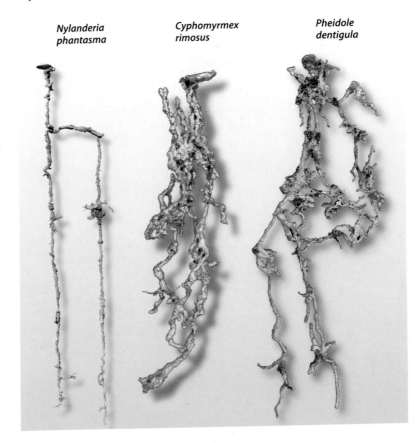

Nylanderia
phantasma

Cyphomyrmex
rimosus

Pheidole
dentigula

FIG. 9.12. Loss of distinct
chambers occurs in several
species. Author's photos, in
part from Tschinkel (2015a).

those of fire ants. Once again, the multiplication of modules is the basic process underlying this evolution.

A final change from the ancestral nest organization (fig. 9.1) is the complete
loss of chambers, so that all that remains is a vague widening of the shaft here
and there as in *Nylanderia arenivaga* (fig. 9.12), although near-surface chambers
sometimes remain. The extreme of this trend combines with branching to produce nests that appear unorganized, almost chaotic, a haphazard web of shafts
as in *Cyphomyrmex rimosus* and *Pheidole dentigula* (fig. 9.12). Finally, the workers
of *Dolichoderus mariae* simply excavate all the sand beneath a clump of grass to
expose a scaffolding of roots on which they hang themselves and their brood.
What they create is more like a foxhole than a house (fig. 9.13).

Fig. 9.13. The nest of Mary's tongue-and-groove ant, *Dolichoderus mariae*, is more like a foxhole than a house. During the warm season, this multiple-queen colony expands into many such temporary nests and then retracts to overwinter in only one or two of them. Author's photo, from Tschinkel (2015a).

Superimposed on these trends is the effect of worker body size. The many-fold variation of worker body size across species is accompanied by a parallel variation of total nest size independent of colony size. Figure 9.14 shows nests of two species whose colonies have similar numbers of workers but whose workers vary in body size by 25 to 40 times. The large, chunky nest of *Camponotus socius* contrasts with the slender wraith of a nest of *Pheidole adrianoi*. Similar effects of body size are expressed in many ant species.

As with the evolution of chambers, we can propose a hypothesis for the evolution of different arrangements of chambers and shafts within the nest (fig. 9.15). The schematic is intended only to show how states can change. Different sequences of changes may arrive at the same end state, but unfortunately the sample of species is far too small to make reliable claims for sequences of evolution. As with chambers, many of these changes seem to be able to evolve independently of each other, giving the characters a mosaic distribution across

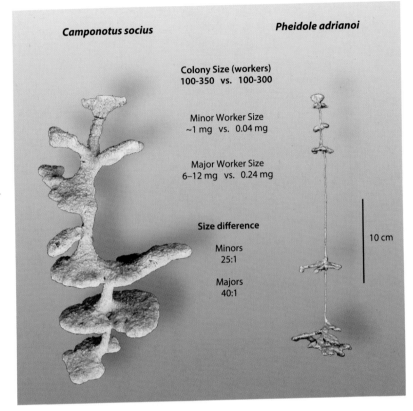

Camponotus socius

Pheidole adrianoi

Colony Size (workers)
100-350 vs. 100-300

Minor Worker Size
~1 mg vs. 0.04 mg

Major Worker Size
6–12 mg vs. 0.24 mg

Size difference

Minors
25:1

Majors
40:1

10 cm

FIG. 9.14. The effect of worker body size can be easily recognized when nests are excavated by approximately the same number of workers. Colonies of both *Camponotus socius* and *Pheidole adrianoi* range up to a few hundred workers, but workers of *C. socius* are 25 to 40 times as heavy (minors, majors) as those of *Ph. adrianoi*. Both nests are to the same scale. Author's photos.

species. Not all the states shown in figure 9.15 have been discovered in my research, but this doesn't mean they don't exist.

HOW LABILE IS NEST ARCHITECTURE?

There seems to be a lot of "mix and match" in the architecture of ant nests, by which I mean that the chambers and shafts that make up the nests seem to occur in many combinations, as though they evolved independently of one another. This raises the centrally important question of just how easily evolution can change these nest characteristics. Are some combinations of characters more conserved than others—that is, do they tend to evolve linked together? Asked another way, do particular lines of ant evolution build nests that are similar within the line but different from those of other lines? Or are such patterns

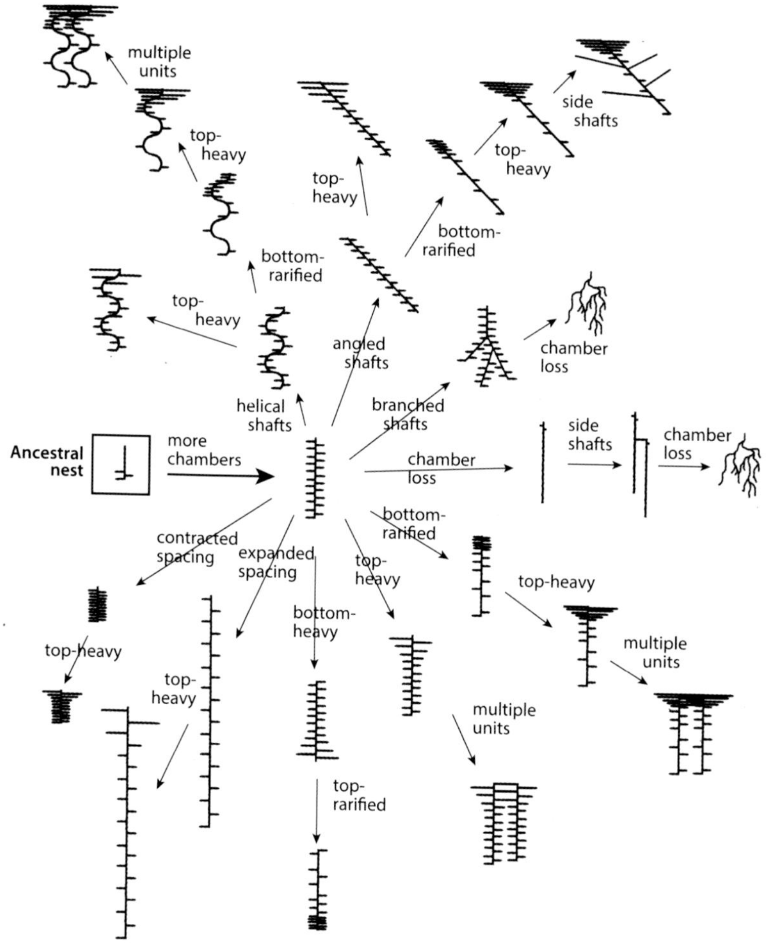

FIG. 9.15. Evolutionary and developmental changes in chamber arrangement that lead to a range of nest types.

more or less absent? In other words, is the similarity of architecture correlated with the relatedness of the ants? Answering this question requires large samples that can be arranged into phylogenetic trees. While this has been done for ant species based on their morphological and genetic characters, I am currently not able to answer this question for nest architectures, especially not for those of distantly related species. My sample of casts is simply too small and too limited to only some groups of ants. The 43 species of which I have made nest casts (table 9.1) are in only 4 of the 17 subfamilies of ants, and 26 of the species

TABLE 9.1. Species-typical characteristics of nest architecture (species are alphabetical by scientific name)

NO.	SPECIES	FIGURE	CHAMBERS	SHAFTS	OTHER
1	Ashmead's long-legged ant (*Aphaenogaster ashmeadi*)	4.3, 9.17	Simple and few, not very distinguished	Simple	Shallow nests
2	Cockerell's long-legged ant (*Aphaenogaster cockerelli*)	Not pictured	Large, star shaped upper, simpler below	Vertical, robust	
3	Florida long-legged ant (*Aphaenogaster floridana*)	4.2, 9.4, 9.17	3–8 well-spaced, mostly simple chambers	Vertical, often connected to chamber centers	Upper levels connected by multiple shafts
4	Treat's long-legged ant (*Aphaenogaster treatae*)	9.17	Simple, inelegant	Vertical, simple	
5	Florida carpenter ant (*Camponotus floridanus*)	Not pictured	Irregular, messy	Shallow, unclear	Colony has multiple nests
6	Sandhill carpenter ant (*Camponotus socius*)	9.16	Robust and elegant, often lobed	Wandering, thick	Colonies expand into multiple nests in summer
7	Immigrant little fungus gardener (*Cyphomyrmex rimosus*)	9.12	Chamberless	Chaotic, poorly organized	Chambers and shafts not distinguishable
8	Mary's tongue-and-groove ant (*Dolichoderus mariae*)	9.13	Single large chamber under grass clumps	None	Expands into multiple nests in summer
9	Hump-backed cone ant (*Dorymyrmex bossutus*)	9.8	Tiny, distantly spaced	Very deep, straight	Complex, star-shaped uppermost chamber
10	Buren's cone ant (*Dorymyrmex bureni*)	9.16	Tiny, distantly spaced	Very deep, straight	Complex, star-shaped uppermost chamber
11	Crazy cone ant (*Dorymyrmex insana*)	Not pictured	Tiny, distantly spaced	Wandering	Complex, star-shaped uppermost chamber
12	Brown ectatomma (*Ectatomma brunneum*)	Not pictured	Simple to lobed	Straight, not deep	Amazonian species (Peru)
13	Frosty asbestos ants (*Forelius pruinosus*)	Not pictured	Fingered, down-turned edges, clusters of shallow chambers	Multiple, with multiple orientations	Multiple queens and nests, complex
14	Archbold's fleet ant (*Formica archboldi*)	9.18	Thick, dumpy	Multiple, simple	

No.	Species	Figure	Chambers	Shafts	Notes
15	Wily fleet ant (*Formica dolosa*)	9.6, 9.18	Lumpy, dumpy, thick	Branching, angled	Large nests are massive
16	Variable fleet ant (*Formica pallidefulva*)	9.18	Lumpy, dumpy, thick	Single, angled	Similar to *Formica dolosa*
17	Argentine ant (*Linepithema humile*)	Not pictured	Indistinct	Chaotic, interconnected	Multiple queens
18	Metallic trailing ant (*Monomorium viridum*)	9.8	Lobed, very tightly spaced	Single, vertical	Shallow nest, a work of art
19	Kennedy's honeypot ant (*Myrmecocystus kennedyi*)	9.21	Branched and lobed, large	Wandering	
20	Gloomy honeypot ant (*Myrmecocystus lugubris*)	Not pictured	Oval, regular	Wandering	
21	Navajo honeypot ant (*Myrmecocystus navajo*)	9.21	Lobed, robust, star-shaped near surface	Mostly vertical	
22	Sand-loving crazy ant (*Nylanderia arenivaga*)	9.20	Chamberless	2–5 ragged, vertical shafts, only one entrance	Parallel shafts with a single horizontal connection
23	Northern crazy ant (*Nylanderia parvula*)	9.20	Simple, small, variable	Wandering	
24	Ghostly crazy ant (*Nylanderia phantasma*)	9.12	Chamberless	Ragged, vertical	Parallel shafts with a single horizontal connection
25	Southeastern trap-jaw ant (*Odontomachus brunneus*)	9.16	Simple, robust, ragged	Vertical, simple, somewhat wandering	
26	Rosemary big-headed ant (*Pheidole adrianoi*)	4.12, 9.19	Elongate, star shaped, 3–5 levels	Very thin, single, straight	Artful simplicity in a tiny space
27	Bearded big-headed ant (*Pheidole barbata*)	9.6, 9.19	Thin, large, extensive	Direction changing	Large colonies
28	Versatile big-headed ant (*Pheidole dentata*)	9.19	Often indistinct, poorly defined	Some branching, direction changing	
29	Woodland big-headed ant (*Pheidole dentigula*)	9.12, 9.19	Chamberless	Chaotic, interconnected	
30	Morris's big-headed ant (*Pheidole morrisi*)	9.10, 9.19	Many simple, small lateral chambers	Up to 4 flattened shafts	Multiple vertical shish-kebab units

Continued on next page

TABLE 9.1. (*continued*)

NO.	SPECIES	FIGURE	CHAMBERS	SHAFTS	OTHER
31	Large imported big-headed ant (*Pheidole obscurithorax*)	9.19	Mostly simple or somewhat lobed	Single, more or less vertical, very deep	Steplike arrangement of upper chambers
32	Sand-loving big-headed ant (*Pheidole psammophila*)	9.19	Many near surface, small below	Angled to depth	
33	Rough big-headed ant (*Pheidole rugulosa*)	9.19	Small, variable	Wandering	
34	Dry-loving big-headed ant (*Pheidole xerophila*)	9.19	Small, lobed	Wandering	
35	Florida harvester ant (*Pogonomyrmex badius*)	2.2, 4.8–4.11, 9.5, 9.8	Lobed to complex chambers on outer bends, decreased size with depth	Helical	Large, complex chambers near surface, Queen of Nest Architects
36	California harvester ant (*Pogonomyrmex californicus*)	9.22	Large, lobed, complex near surface	Horizontal, then wandering to depth	Very large nests
37	Little harvester ant (*Pogonomyrmex magnacanthus*)	4.1, 9.23	Regular, simple, evenly spaced on both sides of shaft	Branched, rambling, wandering to depth	Very extensive nests
38	Winter ant (*Prenolepis imparis*)	10.1	None shallower than 1 m, highly lobed, well spaced	Extremely deep, single, straight	Tangled tunnels just under the surface
39	Fire ants (*Solenopsis invicta, S. geminata*)	2.1, 9.10, 9.16	Hundreds of small chambers, often merged	Multiple, densely placed, deeper in center of nest	Multiple interlacing shish-kebab units
40	Pergande's thief ant (*Solenopsis pergandei*)	Not pictured	Small, oval	Rambling network in a large volume	
41	Sand-loving creeper ant (*Temnothorax texanus*)	Not pictured	Tiny, lumpy, single	No obvious shaft	Very shallow
42	Tuberculate fungus gardener (*Trachymyrmex septentrionalis*)	7.7, 9.4	1–4 egg-shaped fungus chambers	Simple, vertical	Grows fungus on caterpillar droppings and plant bits
43	Desert black harvester ant (*Veromessor pergandei*)	9.9	Large, flat	Single, angled	Very large, deep nests

are in a single subfamily, the Myrmicinae, with 1 to 9 species in 3 other sub-families (Formicinae, Dolichoderinae, Ponerinae). Published accounts other than mine (not included in table 9.1) expand our knowledge mainly of the fungus-gardening ants, which also belong in the Myrmicinae (e.g., fig. 9.2). References to this literature can be found in the references section. Clearly, I cannot say much about broad patterns across distantly related taxa.

Deducing patterns of evolution across species requires that we compare homologous characters. The more confident we are that particular charac-ters are homologous, the more confident we can be in the construction of phylogenetic trees. Comparing the architecture of modern nests with that of the ancestral nest, we can see that most shafts of modern ant nests must be homologous—that is, all descended from the same ancestral shaft of the ancestral nest. Similarly, most horizontal chambers of modern ant nests must be homologous for the same reason—chamber horizontality is an ancestral trait. Of course, the actual homologues are not shafts and chambers but the behavioral programs of worker ants that produce these structures.

It is much more difficult and uncertain to decide which particular *features* of shafts and chambers are homologous. Are the strong lobes of a chamber in spe-cies *A* homologous to those in species *B*? If the two species are reasonably closely related, say in the same genus, we would be inclined to say they are. We would reason similarly with the arrangement of chambers on shafts. How-ever, it is easy to imagine that small behavioral changes in the workers would independently result in similar lobes or chamber arrangements in distantly re-lated species, or different ones within closely related species (fig. 9.3). In gen-eral, positive feedbacks could amplify small behavioral changes to result in large changes in nest architecture, even in closely related species. As an example of a positive feedback, a small increase in the likelihood that a worker will dig where another has already dug, coupled with an attraction to active workers, will result in a large and focused increase in excavation because each worker's action stimulates action by multiple other workers until crowding limits ac-cess to the working face. The growing cavity will be widened when some of the stimulated workers then move to the periphery, and new lobes will form where a separate group of workers dig. Similar dynamics could result in a variety of chamber shapes.

Of course, any such exercise would be pointless if there were no consistent architectures at the lowest taxonomic level, the species. Thus we must ask whether there is obvious consistency (or not) within species (i.e., is architecture species-typical?). To do this, we must identify distinctive architectural features or combinations of features (fig. 9.15) that may be shared within a species but differ between species. Figure 9.16 shows representative examples of 17 species for which I have made multiple casts and can vouch that nests within the species follow the same basic plan. This figure demonstrates many of the distinctive features of each that allow us to recognize that a nest is the work of a particular species. Table 9.1 briefly summarizes the distinctive features and is meant to work with figure 9.16.

At the next higher taxonomic level, between the related species within a genus, the picture is not so simple. For example, the uniting features of the three species of long-legged ants, *Aphaenogaster*, are less obvious (fig. 9.17). True, all have mostly a single-shaft structure, but the smudgy, sloppy nests of *A. treatae* and *A. ashmeadi* are like trailer trash to the precise, neat elegance of *A. floridana*. Among the three species of the fleet *Formica*, the common feature seems to be thick shafts and dumpy chambers, perhaps because these are all large ants (fig. 9.18). What separates them is the more or less single vertical shaft in *F. archboldi*, the single angled shaft in *F. pallidefulva*, and the multiple angled, branching shafts in *F. dolosa*. As a result, *F. dolosa* has more chamber volume deep (bottom heavy), while the other two are more top heavy. The implied behavioral differences among excavating workers of these three species are easily deduced and hint at how nest architectures probably evolve as changes in the programs that influence how, where, and how much workers excavate. Within the fire ants *Solenopsis geminata* and *S. invicta*, nest architecture is very similar. Chambers are larger and "flatter" in *S. geminata* than in *S. invicta* (fig. 9.10, *left*), and nests of the former are smaller than those of the latter, an obvious result of differences in colony size.

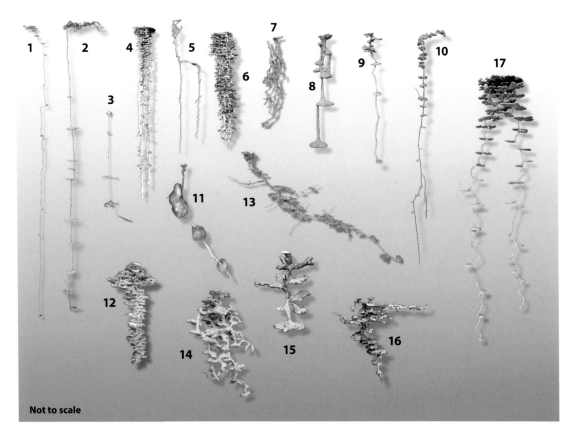

Fig. 9.16. Representative examples of species-typical nest architectures, as judged from multiple casts of each species. (1) *Dorymyrmex bureni*; (2) *Prenolepis imparis*; (3) *Pheidole adrianoi*; (4) *Pheidole morrisi*; (5) *Pheidole dentata*; (6) *Solenopsis invicta*; (7) *Cyphomyrmex rimosus*; (8) *Aphaenogster floridana*; (9) *Odontomachus brunneus*; (10) *Pheidole obscurithorax*; (11) *Trachymyrmex septentrionalis*; (12) *Monomorium viridum*; (13) *Veromessor pergandei*; (14) *Formica dolosa*; (15) *Camponotus socius*; (16) *Formica pallidefulva*; (17) *Pogonomyrmex badius*. Not all cast species are pictured here. Author's photos, in part from Tschinkel (2015a).

But there is plenty of evidence of high lability in nest architecture, for large differences often occur between even closely related species. The big-headed ant genus, *Pheidole*, is one of the largest and most diverse ant genera in the world. My sample of nine North American species suggests that many architectural features are quite labile even within this genus (fig. 9.19). Even aside from the

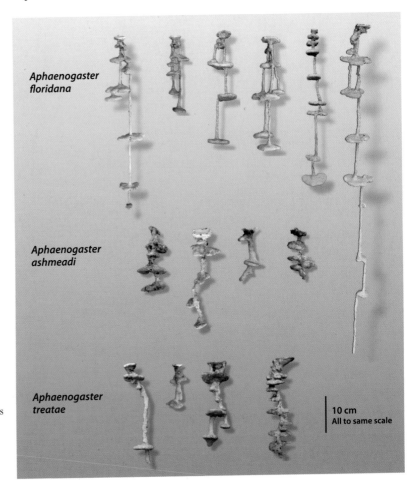

*Aphaenogaster
floridana*

*Aphaenogaster
ashmeadi*

*Aphaenogaster
treatae*

10 cm
All to same scale

Fig. 9.17. Variation of nest architecture among three species within the genus of the long-legged ants, *Aphaenogaster*. Author's photos, modified from Tschinkel (2015a).

size difference of the ants themselves, the delicacy of *Ph. adrianoi* seems a world apart from the dense, uncompromisingly serial nature of *Ph. morrisi* and *Ph. obscurithorax*. The chaotic *Ph. dentigula* and the thin pancakes of *Ph. barbata* seem so different from the aforementioned two species, while *Ph. psammophila* seems to have taken the stepwise upper chambers of *Ph. obscurithorax* as the principle of its entire nest. All this is the more mysterious because *Ph. obscurithorax* is a South American species only distantly related to the species with which its nest shares some features. The odd cast of *Ph. dentata* seems only to compound the confusion.

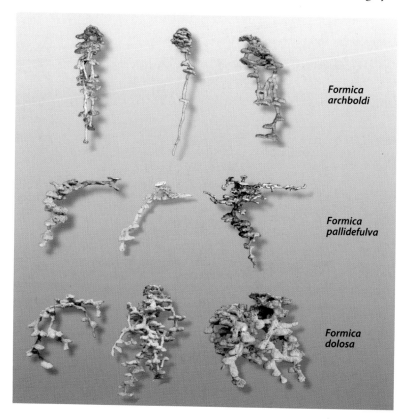

Formica archboldi

Formica pallidefulva

Formica dolosa

FIG. 9.18. Variation of nest architecture among three species within the genus of fleet ants, *Formica*. Author's photos, modified from Tschinkel (2015a).

A similar picture emerges from the three species of the crazy ant *Nylanderia* (fig. 9.20)—there seems little in common between the undistinctive, simple nests of *N. parvula* and the chamberless, ragged nests of *N. arenivaga* and *N. phantasma*. Then, deciding what the two species of *Myrmecocystus* have in common is a head-scratcher (fig. 9.21). In *M. kennedyi* we see a wandering shaft with thin, elongate, branching chambers, whereas in *M. navajo* we see a vertical shaft, star-shaped upper tunnels, and fairly ordinary, simple chambers. What they do have in common is a "high ceiling" in some of their chambers—this is where the replete honey pots hang from the ceiling, waiting to regurgitate their stored sweet (or yucky) fluid during lean times. These nests seem only distantly related, like some of those in the genus *Pheidole*, and once again suggest a high degree of lability in nest architecture among closely related ants.

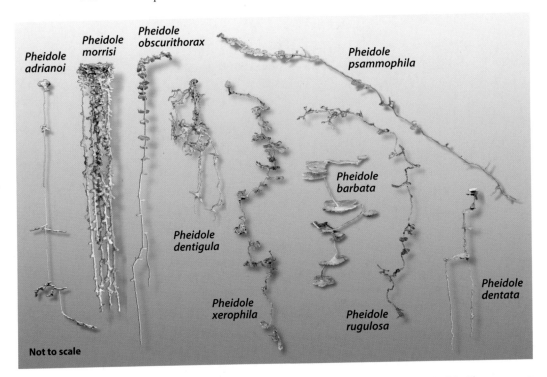

Fig. 9.19. Variation of nest architecture within the genus of big-headed ants, *Pheidole*. Photos are not to the same scale, but *Ph. adrianoi* is the smallest and *Ph. obscurithorax* the largest. Author's photos.

A similar divergence in architecture seems to be present in the three species of *Pogonomyrmex* that I have been able to cast. The Americas are home to dozens of species of *Pogonomyrmex*, and most of them harvest seeds to some degree. In addition to *P. badius*, I have made casts of two species of *Pogonomyrmex* from the desert of Southern California, and both seem inspired by a different school of architecture. *P. californicus*, like *P. badius*, makes large chambers just under the surface, but these chambers are more like slender, radiating tunnels rather than the weblike structures of *P. badius* (figs. 9.5, 9.22). There is no neat, helical descending shaft, but rather a cluster of somewhat deeper chambers directly beneath the nest entrance. These connect through a long, chamberless tunnel to chambers emanating from both sides of a

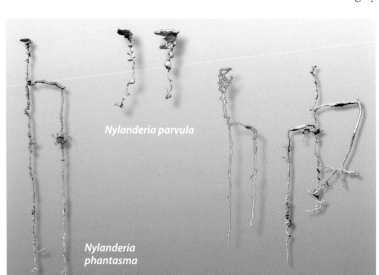

Nylanderia parvula

Nylanderia phantasma

Nylanderia arenivaga

FIG. 9.20. Variation of nest architecture within the genus of crazy ants, *Nylanderia*. Author's photos.

gradually descending shaft. There is beauty here, but not the neat compactness of *P. badius*.

But neither *P. badius* nor *P. californicus* prepares us for the nests of *P. magnacanthus* (fig. 9.23). Gone are the large, complex subsurface chambers; gone is the helix; gone is the gradual change of chamber size and shape with depth. Instead, there is the formicine equivalent of an endless strip mall along an endless highway. One of the branches of the long, branching, shallow subsurface tunnel descends with increasing steepness to great depth. At 2.2 m, my capacity to cast and dig in the dry desert soil finally reached its limit. Chambers were arranged on both sides of these enormously long tunnels and shafts, but never exactly opposite one another. No one would have guessed the extent of these nests from the small handful of soil piled around the inconspicuous nest entrance and the small size of the workers.

That three species within the same genus should make such different nests simply whets the appetite to find out what other *Pogonomyrmex* species do. There are hints from the literature that some of the features of *P. badius* are widespread in the genus, but until I or someone else makes proper nest casts, the

FIG. 9.21. Variation of nest architecture between two species of honeypot ants: *left*, *Myrmecocystus kennedyi*; *right*, *M. navajo*. Author's photos.

answer will be unknown. What already seems clear is that inspired architecture, however that arises among ants, seems widely shared among the species of *Pogonomyrmex*. What is also again clear is that evolution can create large changes in nest architecture, even within closely related species. A few twitches to worker behavior and voilà, different architecture.

Interestingly, it seems likely that *within* a species, the architecture may remain stable for long periods. Jon J. Smith and his colleagues have described

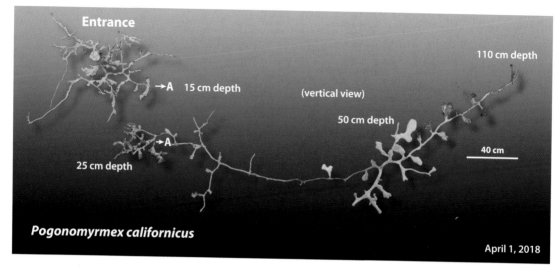

Fɪɢ. 9.22. A nest cast of the California harvester ant, *Pogonomyrmex californicus*. The complex uppermost chambers (*upper left*) are located over the smaller complex. The junction between the two is indicated by the letter *A*. A long, shallow tunnel connects the superficial chamber at the left with the more sharply descending chambers at the right. The deepest-cast chambers are 110 cm deep, but the nest is much deeper. Author's photo.

what appear to be large numbers of fossilized ant nests in the Neogene Ogallala Formation in western Kansas. These nests are 3 to 20 million years old and are similar to those of some modern species, including *Pogonomyrmex* harvester ants, suggesting little change over many millions of years. Sadly, fossilized specimens of the ants that made these nests are not available, but happily, these authors honored me by naming a new type of fossil ant nest *Daimoniobarax tschinkeli*. Perhaps someday, the ant that made this and other nests will be identified and the long-term stability of architecture (and species) over time will be verified or refuted.

If we compare architectures among all the casts I have made and see these in light of the relatedness and family tree of ants, we find a few near-universal characters such as chambers, shafts, and shish-kebab units, although the details vary widely. Some groups of closely related species have moderately consistent architectures, but others show radical differences, suggesting that architecture need not track relatedness closely.

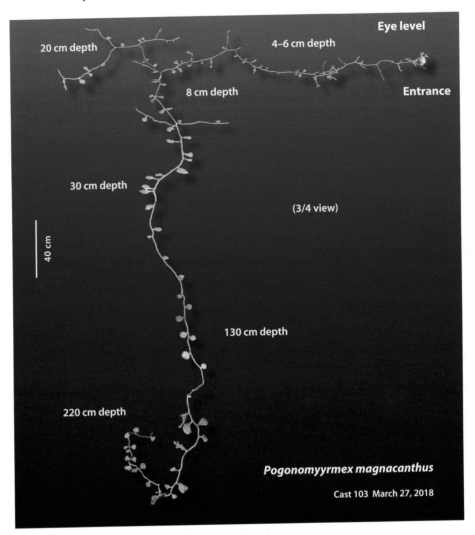

Fig. 9.23. A nest cast of the little harvester ant, *Pogonomyrmex magnacanthus*. A side branch of the very long subsurface tunnel descends to the lowest-cast chamber at 220 cm, but the complete nest is clearly much deeper. In contrast to the nests of *P. badius* and *P. californicus*, this nest is composed of similar repetitions of simple chambers on either side of the shaft, with moderately variable spacing between chambers. The small workers and the handful of sand deposited around the nest entrance give no hint of what lies underground. Author's photo.

Given the high level of variability among nests even within some genera, we have to conclude that the behavioral programs that create the nests are quite flexible and easily modified during evolution. If we look back, this should hardly be a surprise, for we have multiple examples in which the behavioral program that actually plays out varies within a single nest, in relation to not only chamber size but also chamber depth (figs. 9.2, 9.3). Both the shape and size of the chamber and the spacing between chambers vary with depth in several species, suggesting that which "behavioral tape" (or program) is played by the ants' nervous system to produce the behavior is contingent on other factors, and that workers embody multiple program tapes, or perhaps a continuously variable program. Given such contingencies, it should not be surprising that nest excavation behavior is very sensitive to evolutionary modification. How else can we explain the variation we see? The deeper question is, why this variability? Do some architectures serve some functions better than others? So far, as I described in chapter 6, there are few answers to these questions. The frontiers of our quest to explore ant nest architecture are clear, but what lies beyond these frontiers is still a mystery.

CHAPTER 10

Afterword: The Future

At the beginning of this book, I promised a story about the pleasures, challenges, and rewards of low-tech shoestring science. I have now shared this story as far as I have taken it. Along the way, I shared the pleasures of simple science, problem solving, gizmo building, and careful observations, all in intimate contact with nature. I have revealed a world that has been hidden beneath our feet simply by filling the invisible empty spaces of ant nests to make them visible. I have shared the thinking and problem solving it took to develop effective casting methods, and I have used these methods to reveal forms with surprising attraction and beauty, forms about which there is much to wonder—that such small creatures can build something so huge, complex, and beautiful, and do this over a short period in the dark without a blueprint and without a leader. The art of ant colonies with their variations on themes is as aesthetically interesting and pleasing as that of many human sculptors. It is also impressive that ants have evolved this diverse array of nest architectures from only two basic elements: chambers and shafts.

Ant nests now take their place as yet another of nature's infinite forms, each a jewel in its own right. Perhaps a beautiful form stirs emotions much like a beautiful musical passage. In music, particular combinations and timings of notes and harmonies combine to create meaning and pleasure. It seems that just the right combination of planes, curves, and spacing creates a form that sounds a lovely chord in the nervous system, a music of form, shape, texture, and color. And just as variations on musical themes have their special place in the pleasures of music, so do variations of form occupy a special place in the "music of forms." Even within a set of species-typical nests, the inexact rules of construction produce pleasing thematic variations, and among species, these variations can be very large.

But revealing these objects can never be more than a beginning, for the objects cry out for an explanation of how they are created, and why they have their particular forms. Such questions led us into considering ant colonies as superorganisms composed of multiple parts structured at multiple levels in order to carry out all of life's functions, with the ultimate goal of producing more superorganisms like themselves. This structure-function relationship consists of the arrangement of workers of different ages and jobs, stored seeds, and brood within the nest. We can now see the nest as the metaphorical "physical body" of the superorganism, with different parts in different locations carrying out different essential life functions, much like the arrangement of organs within an animal, and with the size of these parts appropriate for the life stage and season. In order to fulfill these functions, an endless stream of ants is born in the bottom of the nest, moves upward, changes jobs, and eventually dies foraging or defending the colony's territory. This is the beauty of function, of being adapted, of persevering against the world's challenges, sometimes winning and thriving, sometimes losing and declining.

We tend to assume that even the details of this variation are functional, for the association of structure and function is a fundamental premise in biology— that for every function there is a structure that carries it out, and (less certainly) every structure has a function. In view of this belief, it is frustrating that we do not know the reason for this diversity. Our ignorance should motivate future research, but what directions might it take? What questions should we seek to answer, and how can they be answered?

LEARNING FROM OTHER FIELDS

In a sense, the question of the function of particular architectural features parallels the question of why plants have so many different leaf shapes and leaf arrangements (architectures). This view is not as out of the blue as it may seem, for ant colonies, ant nests, and plants are all modular entities, composed of repetitive similar pieces (workers; chambers and shafts; leaves and stems). All grow (or shrink) by the addition or loss of modules. I have described the near impossibility of experiments testing particular nest architectural features. Similar challenges would seem to face experiments on particular leaf shapes and arrangements, but botanists have made progress through nonexperimental approaches, most commonly by correlating plant architecture with climatic and other habitat attributes, always in light of evolutionary lineage. Combining such correlations with modeling, they have shown that energy inputs and outputs, heat and cold damage, and the difference between leaf and air temperature affect maximum leaf size. Thus, large-leaved species are more common in the hot, sunny, wet tropics where water for evaporative cooling is not limiting, whereas smaller-leaved species are found more in hot, arid regions where water for cooling is limiting, and in colder regions where cold damage is a threat. Data on leaf sizes and shapes are available for thousands of plant species in almost every conceivable habitat, which is why abundant and powerful correlations can be made for plants.

Similar approaches would seem to have promise for teasing apart the factors affecting nest architecture. But before it will be possible to seek informative correlations, the database needs to be enormously expanded by accumulating the architectures of hundreds more ant species from around the world in a very large range of habitats and soils. The current database consists of dozens of species from a small handful of habitats, not thousands from around the world. With a larger database, it would be possible to ask questions such as, Do nest architectures in sandy soils or cold climates or arid regions or wet tropics or high elevations share certain features? As a small example, it may or may not be meaningful that the shafts of several species I cast in the desert of Southern California are inclined, not vertical, whereas that is rarely the case in Florida. But with only a few casts to compare, all conclusions are on hold.

METHOD IMPROVEMENTS

There are method issues, too. In my studies, I have judged similarity of shapes almost entirely by eye, a sort of "good enough for government work" approach, but for more rigorous comparisons, shapes and whole architectures would really have to be rendered in mathematical or geometric terms that could be quantitatively compared. Such capabilities would allow comparisons across the works of multiple authors without having the actual casts in hand. These methods would also quantify variation within a species and across species, as well as provide rigor to claims that selected architectures are or are not different. They could answer the question of whether the architecture changes (or does not) as colonies grow. I have claimed that it does not, but my methods are rather crude.

FUTURE CHALLENGES

Should architectural similarities among species within a group be discovered, there is still the question of whether these similarities are associated with evolutionary lines of species or arose independently of lineage in response to environmental selection. I have tried to ask this question of my very limited collection of species within genera. The mixed answer is somewhat perplexing, for I found very similar architectures within a genus as well as some very large differences, suggesting that under some evolutionary circumstances, external selective pressures trump heritage within lineages, but in others they are conserved. This question, like many others, awaits a much larger data set.

Geographic variation of architecture within a single species with a large geographic range or a large range of habitat or soil types could point to climatic or soil attributes that affect nest architecture. Many ant species have very large ranges that would be suitable targets for such studies. As an example from widely separated regions, in the Florida coastal plain, nests of *Prenolepis imparis* (fig. 10.1) are often 4 m deep and lack any chambers in the top meter, whereas in Ohio and Missouri they are only 2 m deep and have chambers beginning just under the surface. This correlates with differences in seasonal activity and climate, but probably also with differences in soils. It would be especially valuable to compare species that nest in, for example, sand across a wide range of latitudes, because the effect of soil would cancel out. Sadly, my multiple casts of single species came mostly from the same area south of Tallahassee. Equally

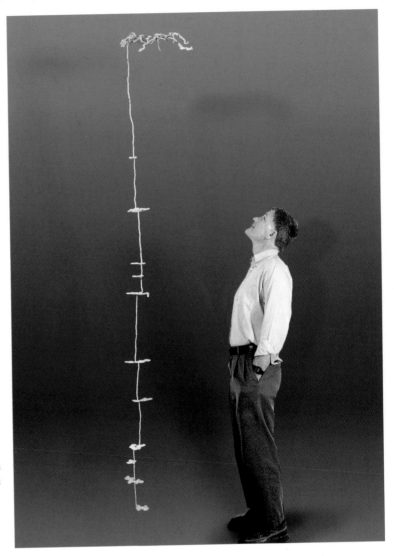

Fig. 10.1. The nest of the winter ant, *Prenolepis imparis*, varies with geographical and climatic differences. In Florida, the nest is very deep and lacks chambers shallower than 1 m (this image), but in Ohio and Missouri, the nest is less than half as deep and has chambers just below the surface. Photo by Charles F. Badland, modified from Tschinkel (2015a).

valuable would be sets taken from different soils within local regions. My single experiment suggested that at least nest size is affected by soil type, but is shape?

The organization of the colony within the nest space may also yield insights. Such information is of course far harder to obtain than a simple cast, but there may be patterns of brood or worker arrangements that correlate with some

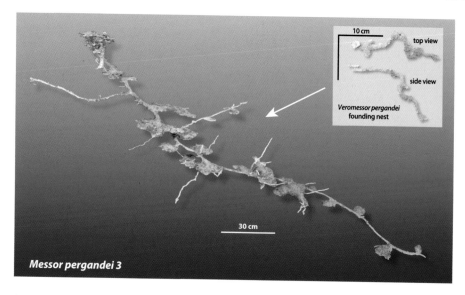

FIG. 10.2. The mature nest is the result of growth from the founding nest dug by the newly mated queen, in this case the black harvester ant, *Veromessor pergandei*. The characteristic inclined shaft so obvious in the mature nest is also present in the founding nest (*inset*) dug by the solitary, newly mated queen, suggesting that queens and workers operate on the same behavioral program. Nests are not to the same scale. Author's photos, modified from Tschinkel (2015a).

kinds of architecture or some of its elements. Other aspects of social organization such as crowding or seasonal differences may also correlate and may provide hints about the functional importance of architectural elements.

In the life cycle of most ants, the nest starts as a tiny founding nest dug by the newly mated queen and is gradually enlarged as her worker population increases during colony growth. This transformation often represents a many thousandfold increase in size, during which, according to my relatively crude methods, the nest's "shape" changes little (figs. 10.2–10.4). It is this lack of change that creates species-typical architecture, no matter what the size, and suggests that the rules of excavation are relative rather than absolute. What these rules are will remain a central question for the foreseeable future.

Eventually it may be possible to hypothesize how physical, social, environmental, and evolutionary factors affect nest architecture and thereby model

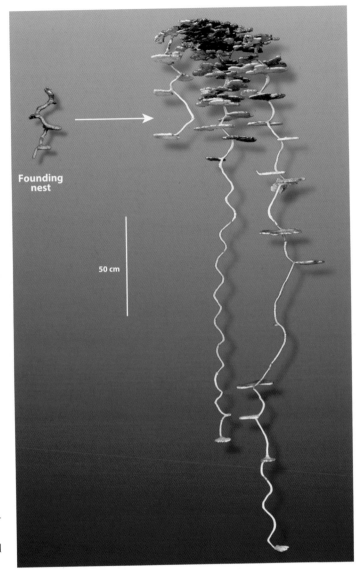

Founding
nest

50 cm

Fɪɢ. 10.3. The mature nest is the result
of growth from the founding nest dug by
the newly mated queen, in this case the
Florida harvester ant, *Pogonomyrmex ba-*
dius. The founding nest (*right*), like the
mature nest, has a helical shaft, suggesting that the founding queen and workers
both operate on the same behavioral program. Nests are not to the same scale.
Author's photos, modified from Tschinkel
(2015a).

potential cause and effect, as has been done for leaf size and shape. I doubt this
will happen soon, but science can't be hurried. When the time is right and the
methods are available . . . one can always hope.

It is obvious that my efforts have barely scratched the surface, so to speak.
Most of the central mysteries of ant nests remain intact. How do workers

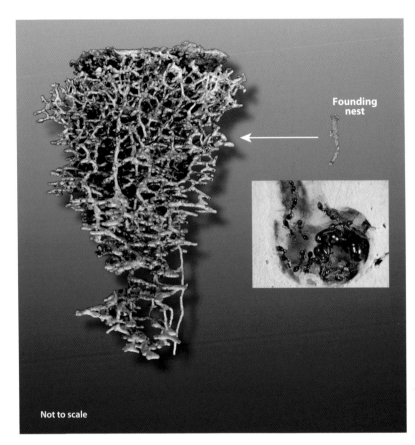

Founding
nest

Not to scale

Fig. 10.4. The mature nest is the result of growth from the founding nest dug by the newly mated queen, in this case the exotic fire ant, *Solenopsis invicta*. The founding nest (*left*) is a simple shaft with a terminal chamber in which the queen rears her first brood (*inset*). Images are not to the same scale. Author's photos, modified from Tschinkel (2015a).

organize themselves to create their nests? How do particular shapes serve particular functions? How does evolution change worker behavior and the resultant nest architecture? How is nest size regulated? What is the full range of nest architecture across the entire family of ants? Are more types of architecture awaiting discovery? All these questions and more await future curious and enterprising biologists. A great deal of work remains to be done, work that will no doubt reveal much that is new, interesting, and beautiful.

SELECTED REFERENCES

Branstetter, M. G., B. N. Danforth, J. P. Pitts, M. W. Gates, R. R. Kula, and S. G. Brady. 2017. Phylogenomic insights into the evolution of stinging wasps and the origins of ants and bees. *Current Biology* 27:1019–25.

Cassill, D. L., and W. R. Tschinkel. 1995. Allocation of liquid food to larvae via trophallaxis in colonies of the fire ant, *Solenopsis invicta*. *Animal Behaviour* 50 (3): 801–13.

Cerquera, L. M., and W. R. Tschinkel. 2010. The nest architecture of the ant *Odontomachus brunneus*. *Journal of Insect Science* 10:64.

Conway, J. R. 2003. Architecture, population size, myrmecophiles, and mites in an excavated nest of the honey pot ant, *Myrmecocystus mendax* Wheeler, in Arizona. *Southwestern Naturalist* 48:449–50.

Cushing, P. E. 1995. Description of the spider *Masoncus pogonophilus* (Araneae, Linyphiidae), a harvester ant myrmecophile. *Journal of Arachnology* 23 (1): 55–59.

Debruyn, L. A. L., and A. J. Conacher. 1994. The bioturbation activity of ants in agricultural and naturally vegetated habitats in semiarid environments. *Australian Journal of Soil Research* 32:555–70.

Diehl-Fleig, E., and E. Diehl. 2007. Nest architecture and colony size of the fungus-growing ant *Mycetophylax simplex* Emery, 1888 (Formicidae, Attini). *Insectes Sociaux* 54 (3): 242–47.

Dlussky, G. M. 1981. *Ants of Deserts*. Moscow: Nauka (in Russian).

Dobzhansky, T. 1973. Nothing in biology makes sense except in the light of evolution. *American Biology Teacher* 35:125–29.

Duarte, A., F. J. Weissing, I. Pen, and L. Keller. 2011. An evolutionary perspective on self-organized division of labor in social insects. *Annual Review of Ecology, Evolution, and Systematics* 42:91–110.

Halfen, A. F., and S. T. Hasiotis. 2010. Neoichnological study of the traces and burrowing behaviors of the western harvester ant *Pogonomyrmex occidentalis* (Insecta: Hymenoptera: Formicidae): Paleopedogenic and paleoecological implications. *Palaios* 25:703–20.

Hamilton, W. D. 1964. The genetical evolution of social behaviour, I. and II. *Journal of Theoretical Biology* 7:1–52.

Harrison, J. S., and J. B. Gentry. 1981. Foraging pattern, colony distribution, and foraging range of the Florida harvester ant, *Pogonomyrmex badius*. *Ecology* 62 (6): 1467–73.

Hart, L. M., and W. R. Tschinkel. 2012. A seasonal natural history of the ant, *Odontomachus brunneus*. *Insectes Sociaux* 59 (1): 45–54.

Hunt, J. H., and C. A. Nalepa, eds. 1994. Nourishment, evolution and insect sociality. In *Nourishment and Evolution in Insect Societies*, 1–19. Boulder, CO: Westview Press.

Jacoby, M. 1935. Erforschung der Struktur des *Atta*-Nestes mit Hilfe des Cementausguss-Verfahrens. *Revista Entomologia* 5:420–24.

Johnson, B. R., and T. A. Linksvayer. 2010. Deconstructing the superorganism: Social physiology, groundplans, and sociogenomics. *Quarterly Review of Biology* 85 (1): 57–79.

Kondoh, M. 1968. Bioeconomic studies on the colony of an ant species, *Formica japonica* Motschulsky. 1. Nest structure and seasonal change of the colony members. *Japanese Journal of Ecology* 18:124–33.

Kwapich, C. L., and W. R. Tschinkel. 2013. Demography, demand, death, and the seasonal allocation of labor in the Florida harvester ant (*Pogonomyrmex badius*). *Behavioral Ecology and Sociobiology*
67 (12): 2011–27.

———. 2016. Limited flexibility and unusual longevity shape forager allocation in the Florida harvester ant (*Pogonomyrmex badius*). *Behavioral Ecology and Sociobiology* 70 (2): 221–35.

Laskis, K. O., and W. R. Tschinkel. 2009. The seasonal natural history of the ant, *Dolichoderus mariae*, in northern Florida. *Journal of Insect Science* 9:2.

MacKay, W. P. 1981. A comparison of the nest phenologies of three species of *Pogonomyrmex* harvester
ants (Hymenoptera: Formicidae). *Psyche* 88 (1–2): 25–74.

———. 1983. Stratification of workers in harvester ant nests (Hymenoptera: Formicidae). *Journal of
the Kansas Entomological Society* 56:538–42.

McGlynn, T. P. 2012. The ecology of nest movement in social insects. *Annual Review of Entomology*
57:291–308.

Mersch, D. P., A. Crespi, and L. Keller. 2013. Tracking individuals shows spatial fidelity is a key regu-
lator of ant social organization. *Science* 340 (6136): 1090–93.

Meysman, F. J. R., J. J. Middelburg, and C. H. R. Heip. 2006. Bioturbation: A fresh look at Darwin's last idea. *Trends in Ecology and Evolution* 21:688–95.

Mikheyev, A. S., and W. R. Tschinkel. 2004. Nest architecture of the ant *Formica pallidefulva*: Structure, costs and rules of excavation. *Insectes Sociaux* 51 (1): 30–36.

Moreira, A. A., L. C. Forti, A. P. P. Andrade, M. A. Boaretto, and J. Lopes. 2004. Nest architecture of *Atta laevigata* (F. Smith, 1858) (Hymenoptera: Formicidae). *Studies on Neotropical Fauna and
Environment* 39:109–16.

Moser, J. C. 2006. Complete excavation and mapping of a Texas leafcutting ant nest. *Annals of the
Entomological Society of America* 99:891–97.

Murdock, T. C., and W. R. Tschinkel. 2015. The life history and seasonal cycle of the ant, *Pheidole
morrisi* Forel, as revealed by wax casting. *Insectes Sociaux* 62 (3): 265–80.

Nicotra, A. B., A. Leigh, K. Boyce, C. S. Niklas, K. J. Jones, D. L. Royer, and H. Tsukaya. 2011. The
evolution and functional significance of leaf shape in the angiosperms. *Functional Plant Biology*
38:535–52.

Nowak, M. A., C. E. Tarnita, and E. O. Wilson. 2010. The evolution of eusociality. *Nature* 466:
1057–62.

Pamminger, T., S. Foitzik, K. C. Kaufmann, N. Schützler, and F. Menzel. 2014. Worker personality
and its association with spatially structured division of labor. *PloS ONE* 9 (1): 8.

Penick, C. A., and W. R. Tschinkel. 2008. Thermoregulatory brood transport in the fire ant, *Solenopsis invicta*. *Insectes Sociaux* 55 (2): 176–82.

Porter, S. D. 1985. *Masoncus* spider: A miniature predator of Collembola in harvester ant colonies.
Psyche 92:145–50.

Porter, S. D., and C. D. Jorgensen. 1981. Foragers of the harvester ant, *Pogonomyrmex owyheei*: A dis-
posable caste? *Behavioral Ecology and Sociobiology* 9:247–56.

Richards, P. 2009. *Aphaenogaster* ants as bioturbators: Impacts on soil and slope processes. *Earth-Science
Reviews* 96:92–106.

Richards, P., and G. S. Humphreys. 2010. Burial and turbulent transport by bioturbation: A 27-year
experiment in southeast Australia. *Earth Surface Processes and Landforms* 35:856–62.

Rink, W. J., J. S. Dunbar, W. R. Tschinkel, C. Kwapich, A. Repp, W. Stanton, and D. K. Thulman.
2013. Subterranean transport and deposition of quartz by ants in sandy sites relevant to age over-
estimation in optical luminescence dating. *Journal of Archaeological Science* 40 (4): 2217–26.

Robinson, G. E. 1992. Regulation of division of labor in insect societies. *Annual Review of Entomology* 37:637–65.

Roth-Nebelsick, A., A. D. Uhl, V. Mosbrugger, and H. Kerp. 2001. Evolution and function of leaf venation architecture: A review. *Annals of Botany* 87:553–66.

Seal, J. N., and W. R. Tschinkel. 2006. Colony productivity of the fungus gardening ant *Trachymyrmex septentrionalis* (Hymenoptera: Formicidae) in a Florida pine forest. *Annals of the Entomological Society of America* 99:673–82.

Seid, M. A., and J. F. A. Traniello. 2006. Age-related repertoire expansion and division of labor in *Pheidole dentata* (Hymenoptera: Formicidae): A new perspective on temporal polyethism and behavioral plasticity in ants. *Behavioral Ecology and Sociobiology* 60 (5): 631–44.

Sendova-Franks, A. B., and N. R. Franks. 1995. Spatial relationships within nests of the ant *Leptothorax unifasciatus* (Latr.) and their implications for the division of labour. *Animal Behaviour* 50 (1): 121–36.

Talbot, M. 1964. Nest structure and flights of the ant *Formica obscuriventris* Mayr. *Animal Behaviour* 12 (1): 154–58.

Tschinkel, W. R. 1987. Seasonal life history and nest architecture of a cold loving ant, *Prenolepis imparis*. *Insectes Sociaux* 34:143–64.

———. 1993. Sociometry and sociogenesis of colonies of the fire ant *Solenopsis invicta* during one annual cycle. *Ecological Monographs* 64 (4): 425–57.

———. 1998. Sociometry and sociogenesis of colonies of the harvester ant, *Pogonomyrmex badius*: Worker characteristics in relation to colony size and season. *Insectes Sociaux* 45 (4): 385–410.

———. 1999a. Sociometry and sociogenesis of colonies of the harvester ant, *Pogonomyrmex badius*: Distribution of workers, brood and seeds within the nest in relation to colony size and season. *Ecological Entomology* 24 (2): 222–37.

———. 1999b. Sociometry and sociogenesis of colony-level attributes of the Florida harvester ant (Hymenoptera: Formicidae). *Annals of the Entomological Society of America* 92 (1): 80–89.

———. 2003. Subterranean ant nests: Trace fossils past and future? *Palaeogeography, Palaeoclimatology, Palaeoecology* 192:321–33.

———. 2004. The nest architecture of the Florida harvester ant, *Pogonomyrmex badius*. *Journal of Insect Science* 4:21.

———. 2005. The nest architecture of the ant, *Camponotus socius*. *Journal of Insect Science* 5:9.

———. 2010. Methods for casting subterranean ant nests. *Journal of Insect Science* 10:88.

———. 2011. The nest architecture of three species of North Florida *Aphaenogaster* ants. *Journal of Insect Science* 11:105.

———. 2013a. Florida harvester ant nest architecture, nest relocation and soil carbon dioxide gradients. *PLoS ONE* 8 (3): e59911.

———. 2013b. A method for using ice to construct subterranean ant nests (Hymenoptera: Formicidae) and other soil cavities. *Myrmecological News* 18:99–102.

———. 2014. Nest relocation and excavation in the Florida harvester ant, *Pogonomyrmex badius*. *PLoS ONE* 9 (11): e112981.

———. 2015a. The architecture of subterranean ant nests: Beauty and mystery underfoot. *Journal of Bioeconomics* 17:271–91.

———. 2015b. Biomantling and bioturbation by colonies of the Florida harvester ant, *Pogonomyrmex badius*. *PLoS ONE* 10 (3): e0120407.

———. 2017a. Do Florida harvester ant colonies (*Pogonomyrmex badius*) have a nest architecture plan? *Ecology* 98:1176–78.

Tschinkel, W. R. 2017b. Lifespan, age, size-specific mortality and dispersion of colonies of the Florida harvester ant, *Pogonomyrmex badius*. *Insectes Sociaux* 64:285–96.

———. 2017c. Testing the effect of a nest architectural feature in the fire ant *Solenopsis invicta* (Hymenoptera:Formicidae). *Myrmecological News* 27:1–5.

Tschinkel, W. R., and D. J. Dominguez. 2017. An illustrated guide to seeds found in nests of the Florida harvester ant, *Pogonomyrmex badius*. *PLoS ONE* 12 (3): e0171419.

Tschinkel, W. R., and N. Hanley. 2017. Vertical organization of the division of labor within nests of the Florida harvester ant, *Pogonomyrmex badius*. *PLoS ONE* 12 (11): e0188630.

Tschinkel, W. R., W. J. Rink, and C. L. Kwapich. 2015. Sequential subterranean transport of sand and seeds by caching in the harvester ant, *Pogonomyrmex badius*. *PloS ONE* 10 (10): e0139922.

Tschinkel, W. R., and J. N. Seal. 2016. Bioturbation by the fungus-gardening ant, *Trachymyrmex septentrionalis*. *PLoS ONE* 11 (7): e0158920.

Wagner, G. P. 1989. The biological homology concept. *Annual Review of Ecology and Systematics* 20:51–69.

Ward, P. S. 2014. The phylogeny and evolution of ants. *Annual Review of Ecology and Systematics* 45:23–43.

Wilkinson, M. T., P. J. Richards, and G. S. Humphreys. 2009. Breaking ground: Pedological, geological, and ecological implications of soil bioturbation. *Earth-Science Reviews* 97:254–69.

Williams, D. F., and C. S. Lofgren. 1988. Nest casting of some ground-dwelling Florida ant species using dental labstone. In *Advances in Myrmecology*, edited by J. C. Trager, 433–44. Leiden, Netherlands: E. J. Brill.

Wilson, D. S., and E. Sober. 1989. Reviving the superorganism. *Journal of Theoretical Biology* 136: 337–56.

Wright, I. J., N. Dong, V. Maire, I. C. Prentice, M. Westoby, S. Díaz, R. V. Gallagher, B. F. Jacobs, R. Kooyman, E. A. Law, M. R. Leishman, Ü. Niinemets, P. B. Reich, L. Sack, R. Villar, H. Wang, and P. Wilf. 2017. Global climatic drivers of leaf size. *Science* 357:917–21.

Yang, A. S. 2007. Thinking outside the embryo: The superorganism as a model for EvoDevo Studies. *Biological Theory* 2:398–408.

SELECTED BOOKS ON ANT BIOLOGY

Choe, J. 2012. *Secret Lives of Ants*. Baltimore: Johns Hopkins University Press.

Hölldobler, B., and E. O. Wilson. 1990. *The Ants*. Cambridge, MA: Harvard University Press.

———. 1994. *Journey to the Ants*. Cambridge, MA: Belknap Press of Harvard University Press.

———. 2009. *The Superorganism: The Beauty, Elegance, and Strangeness of Insect Societies*. New York: W. W. Norton.

Hoyt, E. 1996. *The Earth Dwellers: Adventures in the Land of Ants*. New York: Touchstone.

Keller, L., and É. Gordon. 2009. *The Lives of Ants*. Translated by J. Grieve. Oxford, UK: Oxford University Press.

Moffett, M. W. 2010. *Adventures among Ants*. Berkeley: University of California Press.

Oster, G. F., and E. O. Wilson. 1978. *Caste and Ecology in the Social Insects*. Princeton, NJ: Princeton University Press.

Tschinkel, W. R. 2006. *The Fire Ants*. Cambridge, MA: Harvard University Press.

Wilson, E. O. 1971. *The Insect Societies*. Cambridge, MA: Harvard University Press.

SELECTED REFERENCES ON FUNGUS-GARDENER
NEST ARCHITECTURE

Cardoso, S. R. S., L. C. Forti, N. S. Nagamoto, and R. S. Camargo. 2014. First-year nest growth in the leaf-cutting ants *Atta bisphaerica* and *Atta sexdens rubropilosa*. *Sociobiology* 61 (3): 243–49.

Diehl-Fleig, E., and E. Diehl. 2007. Nest architecture and colony size of the fungus-growing ant *Mycetophylax simplex* Emery, 1888 (Formicidae, Attini). *Insectes Sociaux* 54 (3): 242–47.

Jacoby, M. 1952. Die Erforschung des Nestes der Blattschneider-Ameise *Atta sexdens rubropilosa* Forel (mittels des Ausgußverfahrens in Zement), Teil I. *Zeischrift für Angewandte Entomologie* 34:145–69.

Jonkman, J. C. M. 1980. The external and internal structure and growth of nests of the leaf-cutting ant *Atta vollenweideri* Forel, 1893 (Hym.: Formicidae). Part I. *Zeischrift für Angewandte Entomologie* 89:158–73.

Kleineidam, C., R. Ernst, and F. Roces. 2001. Wind-induced ventilation of the giant nests of the leaf-cutting ant *Atta vollenweideri*. *Naturwissenschaften* 88 (7): 301–5.

Klingenberg, C., C. R. F. Brandão, and W. Engels. 2007. Primitive nest architecture and small monogynous colonies in basal Attini inhabiting sandy beaches of southern Brazil. *Studies on Neotropical Fauna and Environment* 42:121–26.

Moreira, A. A., L. C. Forti, A. P. P. Andrade, M. A. Boaretto, and J. Lopes. 2004. Nest architecture of *Atta laevigata* (F. Smith, 1858) (Hymenoptera: Formicidae). *Studies on Neotropical Fauna and Environment* 39:109–16.

Moreira, A. A., L. C. Forti, M. A. C. Boaretto, A. P. P. Andrade, J. F. S. Lopes, and V. M. Ramos. 2004. External and internal structure of *Atta bisphaerica* Forel (Hymenoptera: Formicidae) nests. *Journal of Applied Entomology* 128 (3): 204–11.

Moser, J. C. 2006. Complete excavation and mapping of a Texas leafcutting ant nest. *Annals of the Entomological Society of America* 99 (5): 891–97.

Rabeling, C., M. Verhaagh, and W. Engels. 2007. Comparative study of nest architecture and colony structure of the fungus-growing ants, *Mycocepurus goeldii* and *M. smithii*. *Journal of Insect Science* 7:40.

Verza, S. S., L. C. Forti, J. F. S. Lopes, and W. O. H. Hughes. 2007. Nest architecture of the leaf-cutting ant *Acromyrmex rugosus rugosus*. *Insectes Sociaux* 54 (4): 303–9.

Wirth R., H. Herz, R. J. Ryel, W. Beyschlag, and B. Hölldobler. 2003. *Herbivory of Leaf-Cutting Ants: A Case Study on* Atta colombica *in the Tropical Rainforest of Panama*. Springer Ecological Studies No. 164. Berlin: Springer-Verlag.

SELECTED REFERENCES ON LABORATORY
AND THEORETICAL STUDIES

Buhl, J., J. Gautrais, J. L. Deneubourg, and G. Theraulaz. 2004. Nest excavation in ants: Group size effects on the size and structure of tunneling networks. *Naturwissenschaften* 91 (12): 602–6.

Deneubourg, J. L., and N. R. Franks. 1995. Collective control without explicit coding: The case of communal nest excavation. *Journal of Insect Behavior* 8:417–32.

Halley, J. D., M. Burd, and P. Wells. 2005. Excavation and architecture of Argentine ant nests. *Insectes Sociaux* 52:350–56.

Theraulaz, G., E. Bonabeau, and J. L. Deneubourg. 1999. The mechanisms and rules of coordinated building in social insects. In *Information Processing in Social Insects*, edited by C. Detrain, J. L. Deneubourg, and J. Pasteels, 309–30. Basel, Switzerland: Birkhäuser Verlag.

Theraulaz, G., J. Gautrais, S. Camazine, and J. L. Deneubourg. 2003. The formation of spatial patterns in social insects: From simple behaviours to complex structures. *Philosophical Transactions of the Royal Society A* 361 (1807): 1263–82.

Toffin, E., D. Di Paolo, A. Campo, C. Detrain, and J. L. Deneubourg. 2009. Shape transition during nest digging in ants. *Proceedings of the National Academy of Sciences of the United States of America* 106 (44): 18616–20.

REFERENCE ON FOSSIL ANT NESTS

Smith, J. J., B. F. Platt, G. A. Ludvigson, and J. R. Thomasson. 2011. Ant-nest ichnofossils in honeycomb calcretes, Neogene Ogallala Formation, High Plains region of western Kansas, U.S.A. *Palaeogeography, Palaeoclimatology, Palaeoecology*. doi:10.1016/j.palaeo.2011.05.046.

SOURCES OF MATERIALS AND SUPPLIES

Boron nitride high-temperature coating: https://www.zypcoatings.com/product/bn-hardcoat-cm/
Ceramic insulating blanket for kilns: https://www.infraredheaters.com/insulati.html
Small foundry supplies: https://smallfoundrysupply.com/
Sources of dental plaster: http://www.atlanticdentalsupply.com/
https://www.darbydental.com/categories/Laboratory/Gypsum--Stone--Plaster-and-Pumice/Castone/8290406
https://www.net32.com/ec/house-brand-yellow-lab-stone-regular-set-d-153263

INDEX

Italicized page numbers refer to figures.

academic field, choosing an, ix

accidents, 21

Acromyrmex, 27

adaptationism, defined, 110

aesthetic interest, x, 2, 3, 51, 57, 133, 206

aging rate of workers, by season, 161–162

air draft for kiln, 18, 20, *20*

alarm pheromone, 34; 4-methyl-3-heptanone as, 34

alleculid beetle larvae, 32–33, *33*, 75

Allegheny mound-building ants, 1

allocation, superorganismal patterns of, 144–145; "organs" of 154; forager regulation, 158–162; overview of, 207; worker distribution within nest, 163–166. *See also* superorganism

allometry, definition of, 64. *See also* scaling

aluminum, alloys of, 18; as casting material, 16, 17, 18, 51

ancestor of ants, 6–7

ancestral ant nest, 7–9, 174–177, *176*; modification of, 177

Anderson, Ralph, xii, 21

ant biology, general, 4

ant colonies, plant-like characteristics, 69

ant distribution within nest, 35, 46. *See also* distribution

ant geographic distribution, 6

ant guests, *33*; as reason for moving, 75

Ant Heaven, description of, 29; mapping populations in, 71; soil dynamics at, 116

ant nest organization, vertical. *See* distribution

ant nests, arboreal, 8; drawings of, *3*; fossilized, 203; theory and modeling of, 27

ant regional abundance, 5–8

ant size comparison, 65–66, *65*

ant species, number of, 5

Antarctic, 46

anticorrosion paint, 20

ants, as agents of bioturbation, biomantling, 116

ants, biology of, 4–8; distribution of, 5–6; diversity of, 5–6; evolution of, 4–7

Aphaenogaster ashmeadi, 50, 196, *198*

Aphaenogaster floridana, 49, 51, 57; comparison within genus, 196, *197*, *198*; nest cast of, *180*, 181

Aphaenogaster treatae, 196, *198*

arboreal ant nests, 8

architectural features, function of, 114, 208; homology of, 195; testing acceptance of, 110

architectural plans, existence of, 96–111

architecture of ant nests, basic questions about, 2; continuous remodeling of, 124; development during life cycle, 211; evolution of through changes in worker behavior, 195; hypothesis for evolution of, 189; lability of, 190; mathematical description of, 209; need for large database, 208; phylogenetic trees of, 191; species specific, 44; species-typical, 196, *197*; variation within genera, 196

aspirator, 31, 41, 55, 56

assembly of plaster casts, 13, 14, *15*

Atlantic Marine Shipyard, 16

Atta laevigata, 27, 177, *178*

Atta vollenweideri, *178*

attributes of the superorganism, 143–144. *See also* superorganism

Australian ant nest architecture, 27, 186, *187*

author, education of, high school and college, ix; graduate school, ix

backfill, 34, 121–122, 125, 130–132

baiting ants to find nests, 53

base station, correction of GPS with, 72

beauty, ix, 15, 60–63, 201, 206–207

behavioral programs, modification of, 205

biological complexity, evolution of, 135–137

biomantling, ants as agents of 115, 116; collective, at Ant Heaven, 125; modeling of, 127, 128; rarity of studies, 133; rate by harvester ants, 127; rate by *Trachymyrmex septentrionalis*, 129

bioturbation, 115, 116, 117, 121, 128–129, 133; ants as agents of 116; collective, at Ant Heaven, 125

birth rate, 45; estimation of, 112

body size, and nest size, 189, *190*; head size as measure of, 49, 147, 149; importance in natural history, 54, 55; metabolic rate in relation to, 64, reproductive output and, 174

boron nitride, 20

brood care, allocation of workers to, 144; and colony growth, 167; and rate of aging, 162; by callow workers, 39, 147–149; by worker size and age, 141–142; communal, 7; connecting foragers with, 163, *165*; efficiency of, 114; flexibility of, 168

brood chamber, 39, *40*, *41*, 166, 168. *See also* chamber

brood rearing efficiency, colony size in relation to, 111; nest subdivision in relation to, 112

brood, location in nest, 44–46

Brown, Grayson, 11

caging colonies, *160*

callow workers, brood care by, 39–44, *44*; cuticle color and, 153, *153*; distribution within nest, 47, 87, 148–149, *165*, *148*; fatness of, 148–149, *148*; wire banding of, 161

Camponotus socius, 45, 54, *180*, 189, *197*

carbon dioxide measurement, 83–84

carbon dioxide soil gradient, as depth cue, 83; antennal sensors for, 84; microbial origin of, 84; production of, 22; removal of, 86; reversal of, 88, *89*; removal, effect on architecture, *86*, 87, *87*

career workers, 142

casting, 11–27; development of, 13–15; early attempts, 11, 18; tiny nests, *64*; with cement, 27

casting material, plaster as, 11–15; metal as, 16–27; search for stronger, 16

catbrier seeds, 34–35

cause and effect, difficulty of establishing, 94; experiments to assign, 94

census of nests, by dissolving or melting casts, 45, 49, *49*, 51; by excavation, 43, 47, 75, 86, 112, 147; for brood production, 113; for sexual production, 172; in layer cake experiment, 130

chamber outlines, tracing, 31, *31*, *32*, 34, 76, 147, 163, 175

chambers, and colony size, 112; brood, *40*, *41*; horizontality of, 62; loss of, 188, *188*; in relation to depth, 59, 60, 61, *181*; placement on shafts, 106–107, *107*, *109*; height and ant size, 181–182, *182*; seed, *35*, *36*; shape and size, creation of, 179, *180*, *183*; size and spacing, evolution of, *180*, 182, 183, *184*, *191*; variation of, 184; vertical size distribution of, 46, *181*

charcoal, as fuel, 18–19, 22–23, 45; on harvester ant disc, 29, 31, 51, 57, 70, *70*, *79*, 126, 154; transport during relocation, 77, 80, 81, 126

chimney, for air draft, 18, 20

coastal plains, forests of the, 2, 5, 10, 49, 51, 128, 209

colonies as superorganisms. *See* superorganism

colony and nest growth, 211

colony as a family, 4, 5

colony census combined with nest architecture, methods for, 48, *49*

colony conflict, as cause of forager mortality, 161

colony life span, 169, *171*

colony size, 1, 58, 80, 112, 144, 147; and mortality rate, 171; and nest size, 190, 196; and sexual production, 173–174; and worker body size, 189; as proxy for fitness, 77; determination of, and resource allocation, 156–157, 162; survival in relation to, 171

color of emitted light, as indicator of temperature, 23

contraptions, for research, ix–x. *See also* gizmos

correlation as causation, problem of, 87, 94

corrosion by molten metal, 20

credit cards, damage by sand, 42

crickets, 75

critical reading, importance of, 114

crowding, xi, 47, 50, 81, 97, 195, 211

crucible, lifting from kiln, pouring aluminum, *24*

crucibles, corrosion of, 20–21, 55, *56*; size and design, 25

Cyphomyrmex rimosus, 188, *188*, *197*

Daimoniobarax tschinkeli, 203

demography, division of labor by, 159, 161–162, 173

dental plaster, for nest casting, 11

depth cues, carbon dioxide as, 83; during nest excavation, 92, 95; experiment testing, 91, *91*, 92, *93*; intrinsic to soil, 108; travel distance as, 61; within nests, 82

description, as basis of research, 3, 4, 143, 144, 147, 150

digging activity, and depth, 91; and worker age, 90; and soil type, 97, *98*; and travel distance, 92

digging behavior, chamber shape resulting from, 179, *179*; worker, 58–59

digging up a live harvester ant nest, 28–44

dime, for scale, *54*, *65*

dissolving plaster casts, 49

distribution, after relocation, 75, *76*; of chamber space, 82, 84, 87, 88, 90, 91, 96, 111, 126–127, *148*, 165; of chamber area, 46, 47, 49; of foragers and transfer workers, 165; of workers within nest, 34, 35, 43, *44*, 48–50, 84, 87, 146–147, 149; of worker-excavators, 179, *179*

distribution, geographic, of ants, 4–6

diurnal activity, 57, 78

diversity of ants, 4, 5, 6, 9, 144, 174

division of labor, 4; by worker size, 141, 159; during colony growth, 142; flexibility of, 168; in ants, 139–143; principles of, 138

Dolichoderus mariae, 188, *189*

Dominguez, Daniel, xii, 43, 72

Dorymyrmex bossutus, 183, 184

Dorymyrmex bureni, *1*, 11, 57, 181, 192

Dunbar, Jim, xiii, 117

Durablanket, 19

East, Christopher, 27, *187*

East, Stephen, 27, *187*

education, of author, high school and college, ix; graduate school, x

evolution, of ants, 6, 9; of sociality, 7

experiments testing architecture effect on fitness, 111–113, *113*

family trees, 6, 9, 174

fatness, worker age and, 149

fire ant, first cast of, 12

fire ants, brood rearing efficiency and colony size, 111–113, *113*

fitness, architecture's effect on, 110; colony size as proxy for, 77; lifetime, in relation to colony size, 172

Florida harvester ant. *See* harvester ant

forager population, estimation of, 152–154; regulation of, 158; turnover of, 155

foragers, initiating nest relocation, 78; life span of, 155; limitation to top of nest, 156

foraging, as source of mortality, 159; colony size and allocation to, 156, 157; seasonal allocation to, 156, *157*

Formica archboldi, 196, *199*

Formica dolosa, available space, 46; distribution within the nest of, 45; nest cast of, 182, *182*, *197*, *199*; surface appearance of, 53; volume per worker, 66

Formica japonica, *3*, 146

Formica pallidefulva, 196, *197*, *199*

Forti, Luiz, 27, 219

founding nest, 68, 71, 133, *211–213*

foxhole, 188

fragility of plaster casts, 13, 15

frequency of moves. *See* relocation

fungus-gardening ants. *See* leaf cutter ants

Furman, Natalie, 30, 147

future research questions, 213

gas, sampling from soil, 85, *85*

geographic range, 5, 57, 134, 209

George, Neal, xii, 72, 77

ghost, of relocated nest, 70, *70*

Gini coefficient, 123

gizmos, ix–x

glass sandwich nest, 99, 125; construction of, 99

Google Earth, ix, 72, 73, *73*

GPS, mapping populations with, 72

granularity of soil, relative to worker size, 68

grave, 42

Haight, Kevin, x; 21, 37

Hanley, Nicholas, xii, 72, 112, 129, 163, 164

haplo-diploid sex determination, 7–8

harvester ants, xiii; acceptance of variations in architecture, 99–107, 111; aluminum casting of nest of, *24*, *212*; as model superorganism, 146; carbonized ants in nest of, *48*, casts of nests of, *180*, 181, 185, *197*, 200, *212;* charcoal on nest of, 51; depth cues and nest architecture, 83, *89*, 90; excavating a live nest of, 27, 28–45; formation of pellets by workers, *58*; fossil nest of, 203; helical shafts in nests of, 185; location of ants within nests, 45–47; marking workers of, 153; movement of soil, collective, by, 117–122, 125, 128; nest attributes of 96–99, 105, 109–110, 185; nest chambers of, 59, *61*, *65*; nest construction and remodeling by 116–125;

harvester ants (*continued*)
  other species of, 203–204, 211; overview of colony biology, 172–174; plaster cast of, 13, 15, *15*; queen of nest architects, 57–58; relocation of nests, 70, *70*, 81, 96; replanting a nest of, 47; rules for creating nest of, 96–99, 110; sexual production and colony size, 172; tracking colonies of, 170
heat loss rate, casting and, 55
heat shields, emergency, *25*
Heath, Sandy, xii, 21
helical shaft. *See* shafts
homology, definition and examples, 175; of architectural features, 195
Howard, Dennis, xii, xiii, 21

ice nest, 47, 100, *100*, 102, 166; chamber shapes of, 108; chamber spacing in, 104; facsimile of natural nest, 99, 102; horizontality of chambers, 104; method for creating, 100, *101*; reversed chamber order, 102, *103*
insulation, of kiln, 19
invention, importance to scientific progress, 4; 55
Iridomyrmex purpureus, 186, *187*

Jacoby, Meinhard, 26
Jeanne, Bob, 114

Kevlar gloves, 19, 25
kiln, construction of, 18–20; early design of, 18; final design, 22; overheating of, 21; red-hot, ready to pour, *23*; with tools, *19*; various sizes of, *20*
kilns, sizes of, 20, 22
King, Joshua, xii, xiii, 19
Kondoh, Masaki, *3*, 146
Kwapich, Christina, xii, xiii, 25; experiments of, 118, 150, 153–161

lability of architecture, 190, 197, 203
labor transitions, by worker age, 141–142, 160, 162, 167–169
larval care. *See* brood care
Laskis, Kristina, xii
law of radiation, 25
layer cake experiment, harvester ants, *118*, 118–119, *119*, 121, *121*, *122*; *Trachymyrmex septentrionalis*, 129–133
leafcutter ants, 1, 2, 27, 51, 58; chamber shape in nests of, 110, 179; modularity of nests of, 177, *178*, 195; nest relocation of, 70; trash workers of, 142

Leon Scrap Metals, 17
Leptothorax muscorum, 5
levels of natural selection, 137
life span, colony, 129, 143, 145, 151, 169, 172, 174; estimation of, 72, 126, 170, *171*
lighting charcoal, 23
list of nest casts, 52, 192–194
Lofgren, Cliff, 11
low-tech science, 4

mapping, of Ant Heaven colony population, 71, *73*
maps, Google Earth, 72, *73*
marking workers, printers ink for, 153; copper wire for, 153, *153*
mark-recapture, 151–153, *153*; assumptions of, 151; as dilution, 152; for estimating worker life span, 155, 159
Masoncus pogonophilus, *33*, 33, 37, 75
melting metal for casting, 19
metal casting, of nests, 16–27
Michener Paradox, 111
mites, *33*
modularity, 177; of ants and plants, 207
Monomorium viridum, 58, 183, *184*, 197
mounds of soil, 1, *1*, 42, 51
moving nest. *See* relocation
multicellular organisms, origin of, 136
Murdock, Tyler, xii, 45, 72
Myrmecocystus kennedyi, *180*, 199, *202*
Myrmecocystus navajo, 199, *202*
myrmecophiles, *33*

nest architecture, early research, 2; function of, 28; specific functions of, 109
nest casts, Table of species, 52, 192–194
nest digging, worker behavior, 9; evolution of, 9,
nest excavation by ants, gradual during relocation, 81; possible mechanism of, 90; rules for, 96
nest, live, excavation procedure, 28–44, *30*, *31*, *32*, *42*
nest relocation. *See* relocation
nest size, and census, 50; and body size, 55; and worker size, 65–66; effect of soil on, 97; in relation to colony size, 97
nest subdivision, relationship to brood rearing, 111–113, *113*
nest volume in relation to depth, 90
nest, craters/discs/ mounds, *1*, 53, *53*

nesting, in trees, 8; in preformed cavities, 69

Newton's law of cooling, 16

nuclear missiles, targeting of using GPS, 73

Nylanderia arenivaga, 56, 188, 199, *201*

Nylanderia parvula, 199, *201*

Nylanderia phantasma, *188*, 199, *201*

Odontomachus brunneus, 45, 65, 193, *197*, 215

optically stimulated luminescence dating (OSL), 117, 121–123

organization of labor, spatial, 143, 145, 146

parallels, organismal and superorganismal, 137, 173

patterns of distribution within nests, 147. *See also* distribution

pellet formation behavior, 58–59, *58*

pellets, sand, 67; alteration during transport, 123; formation of by workers, 58–59, *58*, 120; loss of sand during transport, 125; rapid transport of, 124; size and worker size, 68

Pheidole adrianoi, 54, 56, 63, *64*, *65*, 67, 189, 198, *197*, *200*

Pheidole barbata, 182, *182*, 198, *200*

Pheidole dentata, 198, *200*

Pheidole dentigula, 53, 188, *188*, 198, *200*

Pheidole morrisi, distribution of ants within the nests of, 45–46, nest mounds of, 51, 53, shish-kebab nest units of, 185–186, *186*, 193, 198, *197*, *200*

Pheidole obscurithorax, 198, *197*, *200*

Pheidole psammophila, 198, *200*

Pheidole rugulosa, *200*

Pheidole xerophila, *200*

phone, damaged by sand, 42

phylogenetic trees. *See* family trees

plaster for nest casting, 11, 13

Pogonomyrmex californicus, 200, 203

Pogonomyrmex magnacanthus, 48, 201, 204

Pogonomyrmex badius. *See* harvester ant

population size estimation. *See* mark-recapture

pouring molten metal safely, 25–26, *24*, *26*

preferences, architectural, by ants. *See* architectural plans

Prenolepis imparis, 147, *197*, 209, *210*

problems, solving of, ix, 2, 206

queen, role in colony biology, 4, 7, 8; evolution of, 7; finding the, 41, *41*; founding nests, *211*, 211, 212, *212*, *213*; of nest architects, 57–58; in layer cake experi-

ment, 118, 129; location in nest, 146; life span synonymous with colony life span, 169; role in colony reproduction, 69; role in the superorganism, 137, 139, *140*, 141, 142, 167; seasonal adaptations of, 173

Queen of Nest Architects. *See* harvester ant, *Pogonomyrmex badius*

reconstruction of plaster cast, 13–15, *15*

red ant, 57

refractory insulation blanket, 19

regional abundance, 5–8

relatedness among workers, 8

relative size, ants, 54

relative size, nests and ants, 50, 65–66, 84, 96, 117

relocation, and creation of architecture, 71; causes of, 75; daily and seasonal rhythm of, 74, 78; comparison of old and new nest, 76; cost of, 77; detection of, 70, 72; direction of, 74; reasons for, 76; organization of, 78, 81; progress of, 77, *80*; trail traffic during, 79

replanting colonies, 29, 47

replication of experiments, importance of, 94

research methods, simple, 2–4; pioneering, 3

research site, Ant Heaven, 66

resources, allocation of within superorganisms, 144. *See also* superorganism

Rink, Jack, xiii, 117

Royce, Elliott, xii

sacrificial zinc anodes, 16, *17*

safety, pouring molten metals, 25–26, *24*, *26*

sand, below-ground deposition of, *122*, *126*. *See also* bioturbation

sand, bulk density of, 120; colored, use in experiment, 38, 119, *119*, *122*, *126*. *See also* soil

sand movement. *See* biomantling

scaling, ants and nests, 63, 65; definition of, 64; scale dependence, 68

scuba tanks, xiii, *17*, 18, 21

Seal, Jon, 129

seasonal life history, 46

seed chambers, 35, *36*. *See also* chambers

seeds, 28, as bait for foragers, 154–155; as propagules, 69; as winter stores, 173; collection by harvester ants, 5, 57, 169; distribution of, with nest, *44*, 75, 86, 87, 93, 163, 207; germination of stored, 37; other harvester ants that collect, 200; recovering from harvester ant

seeds (*continued*)
nests, 32, 34–37, 147; species of, 37; springtails on, 75; storage in chambers, 36, *36*, 75, 165, 168; transport during nest relocation, 77, *80*, 81, 82, 119, *165*, 166–167, 169, 200, 207; transport within nest, 163, 165, *165*, 166–168

sex determination, 7–8

sexual brood, location in nest, *41*, 41, 45

sexual production, colony size and, 158, 169, 174, 111–112; in the superorganism, 137–139; lifetime, 172; reduced, as a cost of nest relocation, 77; seasonality of, 157–158, 162

shafts, acceptance of variations of, 105, *106*; angle of descent of, 105, 184; extraneous, 108; helical, 38, 185; placement of chambers on, 106, *107*; shish-kebab units, 185, *186*; variation of, 185

shop vacuum, 28, *29*, 31–35, 41, 156

shovel, 4, 28, 30, 37, 40, 120, 156; Excalibur, the world's greatest, 38, *39*

silverfish, *33*, 35, 75

simulation of collective sand movement. *See* biomantling, bioturbation

size difference, among ant species, 63, 65, 198

social structure, and nest architecture, 28, 149, 175

sociality, evolution of, 4–8

sociogenesis, 145

socks, lighting on fire, 25

soil, as a medium for animals, 9; burial of objects in, 115; deposition underground, 117; discs and craters, 1, 2, 51; excessively drained, 29; fauna, 115; formation of, 115; granularity of, 9, 62, 66, 68, 115, 120; granularity relative to worker size, 66; of coastal plain forest, 10; OSL dating, effects of ants on, 117; rejuvenation of by ants, 133. *See also* bioturbation/biomantling

Solenopsis geminata, 185, 196

Solenopsis invicta, 51, 58, *113*, 185, *186*, 196, *213*; queen and retinue, *140*, *197*; plaster cast of nest, *12*

Solenopsis nickersoni, 53–54, *54*

solving problems, the pleasures of, ix, 2, 206

spatial fidelity of workers, 146

spatial organization, and function, 28; of labor, 167

species, Table of, 52, 192–194

species diversity and size, 37

spectroscopy, ix

springtails, 5, *33*, 36, 37, 75

sting, harvester ant, 32

stories, really good, 76

superorganism, allocation within, 154–155; attributes of, 143–144; evolution of the, 137–139; functions within the, 143–145, 207; harvester ant as model superorganism, 147–150; lifetime fitness of, 172; lifespan of, 169–171; metaphor of, 135; overview of, 167, *168*, 174, 207; regulation of foragers within, 159–162; seasonal adaptations of, 172–173

survival, colony in relation to size, *171*

Table of species, 52, 192–194

tarp, 28, 29, 30, 42

temperature, 16, 18, 19, 21, 44–46, 55, 78, 82, 83, 112, 208; heat seeking by ants, 45

temperature, revealed by emitted light color, 18, 23

tenebrionid beetles, ix–x; 167

theory and modeling of ant nests, 27

thief ants, 10, 53, *54*

tiny nests, finding 51; casting of, 56

Trachymyrmex septentrionalis, 51, 128, 177, 179; bioturbation/biomantling by, *131*, *132*, 129–133, *197*; nest cast, *130*, *180*

tracing chambers, 31, *31*, *32*, 34, 76, 147, 163, 175

transfer workers, definition and marking, 163; experiment on, 163, *164*, *165*; age in relation to, 166

transport of material within nest, seeds and food, 166

transport of materials during relocation, 80; by worker classes during relocation, 82

trash and litter, 29

travel distance, as depth cue. *See* depth cues

Tschinkel, Henry, xii

tunnel and shaft, definition of, 180

Tyvek heat shields, 25

upward migration of aging workers, 167

vacuum, shop, as a tool for excavations, 28, 31, 34, 41

Veromessor pergandei, 184–185, *185*, *197*, 211

vertical distribution of workers by age, 148, 167; reliability of, 149

virtual harvester ant nest video, 43

wallet and phone, 42

water table, 10, 85

wax casts, 45–46, 50–51, *50*

Williams, David, 11
wonder, ix, xii, 47, 206
worker age, spatial organization by, 146, *148, 179.*
    See also distribution
worker aging, slow and fast, 162
worker cuticle color and age, 39, *153*

workers, body size of, and nest architecture, 189,
    190
workers, evolution of, 7

zinc, as casting material, 16, 17, 21, 51, 64, 55, 56; com-
    bustion of, 55, *56*; density of, 17; source of, 16, *17*